ISSN 0044-2267

ZAMM

ZEITSCHRIFT FÜR ANGEWANDTE MATHEMATIK UND MECHANIK

INGENIEURWISSENSCHAFTLICHE FORSCHUNGSARBEITEN

BAND 62 1982 HEFT 1

AUS DEM INHALT

AKADEMIE-VERLAG · BERLIN

ZAMM Bd. 62 Nr. 1 S. 1–64 Berlin, Januar 1982

EVP 18,— M 34115

INHALT

Wir bitten, **Manuskriptsendungen** zweifach (Original und eine Kopie, sprachlich einwandfrei, Formeln mit Maschine oder in Druckschrift geschrieben) an folgende Anschrift zu richten:

Zeitschrift für Angewandte Mathematik und Mechanik z. Hd. Herrn Prof. Dr. Günter Schmidt
Institut für Mechanik der Akademie der Wissenschaften der DDR
DDR-1199 Berlin, Rudower Chaussee 5

Zu den Arbeiten, die als Hauptaufsätze bestimmt sind, ist auf gesondertem Blatt eine Zusammenfassung von 5 bis 10 Zeilen in englischer und (möglichst) deutscher und russischer Sprache beizufügen. Ausführliche Hinweise für die Autoren, um deren strikte Berücksichtigung gebeten wird, finden sich im Anschluß an das Inhaltsverzeichnis des Jahrganges **61** (1981).

Die Autoren erhalten von den Hauptaufsätzen 75, von den Kleinen Mitteilungen 25 Sonderdrucke ohne Berechnung, darüber hinaus weitere Sonderdrucke gegen Berechnung.

Bestellungen sind zu richten

— **in der DDR** an den Postzeitungsvertrieb unter Angabe der Kundennummer des Bestellenden oder an den AKADEMIE-VERLAG, DDR-1086 Berlin, Leipziger Str. 3—4

— **im sozialistischen Ausland** an eine Buchhandlung für fremdsprachige Literatur oder an den zuständigen Postzeitungsvertrieb

— **in der BRD und Berlin (West)** an eine Buchhandlung oder an die Auslieferungsstelle KUNST UND WISSEN, Erich Bieber OHG, D-7000 Stuttgart 1, Wilhelmstraße 4—6

— **in den übrigen westeuropäischen Ländern** an eine Buchhandlung oder an die Auslieferungsstelle KUNST UND WISSEN, Erich Bieber GmbH, CH-8008 Zürich, Dufourstraße 51

— **im übrigen Ausland** an den Internationalen Buch- und Zeitschriftenhandel; den Buchexport, Volkseigener Außenhandelsbetrieb der Deutschen Demokratischen Republik, DDR-7010 Leipzig, Postfach 160; oder an den AKADEMIE-VERLAG, DDR-1086 Berlin, Leipziger Straße 3—4

ZEITSCHRIFT FÜR ANGEWANDTE MATHEMATIK UND MECHANIK
Herausgeber und Chefredakteur: Prof. Dr. Günter Schmidt. Redaktion: Dr. Winfried Heinrich, Dr. Horst Weinert, Dipl.-Math. Friedhild Dudel, Helga Rühl, Institut für Mechanik der Akademie der Wissenschaften der DDR. Verlag: Akademie-Verlag, DDR-1086 Berlin, Leipziger Straße 3—4; Fernruf: 2236119; Telex-Nr.: 114420; Bank: Staatsbank der DDR, Berlin, Kto.-Nr.: 6836-20712. Anschrift der Redaktion: Institut für Mechanik der Akademie der Wissenschaften, DDR-1199 Berlin, Rudower Chaussee 5; Fernruf: 6802841/3639

Veröffentlicht unter der Lizenznummer 1282 des Presseamtes beim Vorsitzenden des Ministerrates der Deutschen Demokratischen Republik. Gesamtherstellung: VEB Druckerei „Thomas Müntzer", 5820 Bad Langensalza. Erscheinungsweise: Die Zeitschrift für Angewandte Mathematik und Mechanik erscheint monatlich. Die 12 Hefte eines Jahres einschließlich Tagungsheft bilden einen Band. Bezugspreis je Band 360,— M zuzüglich Versandspesen (Preis für die DDR 216,— M). Bezugspreis je Heft 30,— M (Preis für die DDR 18,— M). Bestellnummer dieses Heftes: 1009/62/1.

AN (EDV) 35937

ZAMM 62, 1 – 5 (1982)

P. BASSANINI

Iterative Methods for Contraction Operators in the Product of Two Metric Spaces

Es seien (S_1, d_1), (S_2, d_2) vollständige metrische Räume (bzw. abgeschlossene Teilmengen davon), $(S, \tilde{d}) = (S_1 \times S_2, \tilde{d})$ und $T: S \to S$, $X = Tx$, sei ein Lipschitz-stetiger (strikt) kontraktiver Operator in S für passend gewählte Lipschitz-Konstante und $\tilde{d}$. Es werden Bedingungen angegeben, unter denen Iterationsverfahren existieren, deren Rechenaufwand geringer ist als der von $x_n = Tx_{n-1}$.

Let (S_1, d_1), (S_2, d_2) be (closed subsets of) complete metric spaces, $(S, \tilde{d}) = (S_1 \times S_2, \tilde{d})$ and let $T: S \to S$, $X = Tx$ be a Lipschitz continuous operator (strictly) contractive in S for a suitable choice of its Lipschitz constants and of $\tilde{d}$. Conditions are given under which there may exist iteration procedures whose expense of computer time is lower than required by the usual iterative scheme $x_n = Tx_{n-1}$.

Пусть (S_1, d_1), (S_2, d_2) полные метрические пространства (либо замкнутые подмножества таких пространств), $(S, \tilde{d}) = (S_1 \times S_2, \tilde{d})$ и, дальше, пусть $T: S \to S$, $X = Tx$ является непрерывным по Липшицу (строго) контрактивным оператором в S при надлежащем выборе постоянной Липшица и $\tilde{d}$. Даются условия, при которых существуют итерационные методы, обьем вычислений для которых меньше чем для $x_n = Tx_{n-1}$

1. Let (S_1, d_1), (S_2, d_2) be (closed subsets of) complete metric spaces, $S = S_1 \times S_2$, $x = (u, v) \in S$, $u \in S_1$, $v \in S_2$, and let $T: S \to S$, $X = Tx$, $X = (U, V)$, be a map of the form

$$U = T_1(v, V)\, u\,, \qquad V = T_2(u, U)\, v\,.$$

The aim of this paper is to develop iterative methods for the solution of the fixed point equation

$$x = Tx\,, \qquad x \in S\,, \tag{1}$$

by availing ourselves of ideas due to K. R. SCHNEIDER [1] for the sum of two (strictly) contractive operators in a Banach space (see also [2–4]). For the sake of brevity, we shall henceforth denote by d both metrics d_1 (in S_1) and d_2 (in S_2); whether $d = d_1$ or $d = d_2$ will be clear from the text.

Our basic assumptions will be (H1) through (H5) below.

(H1) $T_1: S_2 \times S_2 \times S_1 \to S_1$, and $T_2: S_1 \times S_1 \times S_2 \to S_2$ are *Lipschitz continuous* [2]:

$$d(U, U') = d(T_1(v, V)\, u\,,\ \ T_1(v', V')\, u') \leqq \lambda_1 d(v, v') + \lambda_2 d(V, V') + \lambda_3 d(u, u')\,;$$

$$d(V, V') = d(T_2(u, U)\, v,\ T_2(u', U')\, v') \leqq \lambda_4 d(u, u') + \lambda_5 d(U, U') + \lambda_6 d(v, v')$$

for all $u, u' \in S_1$, $v, v' \in S_2$, corresponding to U, $U' = T_1(v', V')\, u'$ in S_1, V, $V' = T_2(u', U')\, v'$ in S_2, and $\lambda_i \geqq 0$ $(i = 1, \dots, 6)$.

(H2) $\lambda_2 \lambda_5 = 0$, $\lambda_3 \lambda_6 (\lambda_1 + \lambda_2)(\lambda_4 + \lambda_5) > 0$.

We will find conditions under which T is a (strictly) contractive operator in $(S, \tilde{d})$ with contraction constant k, $0 < k < 1$:

$$\tilde{d}(Tx, Tx') \leqq k\tilde{d}(x, x') \tag{2}$$

for all $x = (u, v)$, $x' = (u', v')$ in S, where $(S, \tilde{d})$ is the (complete) product metric space $S_1 \times S_2$ with metric defined by

$$\tilde{d}(x, x') \overset{\text{def}}{=} p[\alpha d(u, u') + (1 - \alpha)\, d(v, v')] \quad \text{for all} \quad x, x' \in S \tag{3}$$

for some $\alpha \in (0, 1)$ to be specified below, and some $p > 0$ (typically, $p = 1$ or $p = 2$).

Problems of this kind arise from several fields of pure and applied mathematics. We give here two examples:

(i) Consider the quasilinear hyperbolic system in bicharacteristic form [5]:

$$\sum_{j=1}^{m} A_{ij}(t, y, z) \left[\frac{\partial z_j}{\partial t} + \sum_{k=1}^{r} \varrho_{ik}(t, y, z) \frac{\partial z_j}{\partial y_k}\right] = f_i(t, y, z)\,, \tag{4}$$

$$i = 1, \dots, m\,, \qquad y = (y_1, \dots, y_r) \in \mathbb{R}^r\,, \qquad t \in (0, a)\,, \qquad a > 0\,, \qquad z = (z_1, \dots, z_m)\,,$$

with boundary data on N arbitrary (not necessarily distinct) hyperplanes a_k, $0 \leqq a_k \leqq a$, $k = 1, \dots, N$, $N \geqq m$ [6, 7, 8]:

$$\sum_{k=1}^{N} \sum_{j=1}^{m} b_{ijk}(y)\, z_j(a_k, y) = \psi_i(y)\,, \qquad y \in \mathbb{R}^r\,, \qquad i = 1, \dots, m\,. \tag{5}$$

Boundary value problems of the form (4), (5) occur in applications, as e.g. in a laser problem in Nonlinear Optics [8], and include the Cauchy problem for $b_{ijk}(y) = \delta_{ij}\delta_{ik}$ and all $a_k = 0$. Then, as shown in [7], problem (4), (5) reduces to the form (1), with S_1 the Banach space of "solutions" $z(t, y)$, and S_2 the Banach space of functions $h(t'; t, y) = g(t', t, y) - y$, where $g(t', t, y)$ yields the (family of) bicharacteristic lines of (4). Here, $S = S_1 \times S_2$ is a Banach space with (weighted 1-) norm corresponding to (3): for other choices of the norm in S cf. [9]. Thus, the results of the present paper apply to quasilinear hyperbolic (integro-differential) systems in diagonal or bicharacteristic form, with Cauchy or boundary data (5).

(ii) Let $\tilde{S}$ be a Hilbert space or a subspace thereof and suppose that $\tilde{S} = S_1 \oplus S_2$, so that every $x \in \tilde{S}$ can be written as $x = u + v$, $u \in S_1$, $v \in S_2$, where S_1, S_2 are (closed) subspaces of $\tilde{S}$. Let P be the orthogonal projection associated to S_1 ($= P\tilde{S}$) so that $Px = u$ and $(I - P)\, x = v$, I the identity operator. Let $E\colon S_E \to \tilde{S}$ be a linear (not necessarily bounded) operator with domain $S_E \subset \tilde{S}$, $\ker E \subset S_1 \subset S_E$, and $N\colon \tilde{S} \to \tilde{S}$ a continuous (not necessarily linear) operator with $\dim (S_1) < \infty$.

Problem: Solve

$$Ex = Nx\,, \qquad x \in S_E\,. \tag{6}$$

Assume there exists a Galerkin-type approximation $u_0 \in S_1$ s.t. $F(u_0) = P(Eu_0 - Nu_0) = 0$ and that the Fréchet derivative $DF(u_0)\colon S_1 \to S_1$ exists and is invertible. Furthermore, assume there exists a partial left inverse of E, i.e. a linear operator $H_2\colon S_2 \to S_2$ such that $H(I - P)\, Ex = (I - P)\, x$ for all $x \in S_E$ [10]. Then the (bifurcation) problem (6) reduces to the form (1) with S_1, S_2 closed subsets of Hilbert spaces, $S = S_1 \times S_2$, and

$$\begin{aligned} U &= T_1(V)\, u = u - [DF(u_0)]^{-1} P[Eu - N(u + V)]\,, \\ V &= T_2(u)\, v = H(I - P)\, N(u + v)\,, \qquad \lambda_1 = \lambda_5 = 0\,, \end{aligned} \tag{7}$$

under suitable additional assumptions (see [10, 11]). Note that the norm in $S = S_1 \times S_2$ defined according to (3): $\|x\|_S = p[\alpha \|u\| + (1 - \alpha)\, \|v\|]$ differs from the original norm of x in $\tilde{S} = S_1 \oplus S_2$: $\|x\|^2 = \|u\|^2 + \|v\|^2$, and $S = S_1 \times S_2$ is **not** a Hilbert space.

Thus, the results of the present paper also apply to bifurcation problems in nonlinear analysis via the alternative (or Ljapunov-Schmidt) method, whose relevance to nonlinear mechanics, particularly in the case "at resonance" ($\ker E \neq 0$) is well-known: we may mention here problems of periodic solutions of nonlinear ordinary and partial differential equations and nonlinear oscillations [10—15].

The present paper completes and simplifies the preliminary analysis (in the Banach space setting) of [9], to which reference will be made for some details and for a numerical example.

2. Combining the inequalities in (H1), and taking (H2) into account yields

$$\begin{aligned} d(U, U') &\leqq k_1 d(v, v') + k_2 d(u, u'); \qquad k_1 = \lambda_1 + \lambda_2\lambda_6\,, \qquad k_2 = \lambda_3 + \lambda_2\lambda_4; \\ d(V, V') &\leqq k_3 d(v, v') + k_4 d(u, u'); \qquad k_3 = \lambda_6 + \lambda_1\lambda_5\,, \qquad k_4 = \lambda_4 + \lambda_3\lambda_5\,. \end{aligned} \tag{8}$$

Besides (H1) and (H2) we assume now (H3):

(H3) $\qquad k_2 < k\,, \qquad k_3 < k\,, \qquad k_1k_4 \leqq (k - k_2)\,(k - k_3)\,.$

Lemma 1: *Under assumptions* (H1)—(H3) *there exists* $\alpha \in (0, 1)$ *such that* $T\colon (S, \tilde{d}) \to (S, \tilde{d})$ *is a contraction with constant* k, *if* $\tilde{d}$ *is defined by* (3).

Proof: For every $x = (u, v)$, $x' = (u', v')$ and corresponding $X = (U, V)$, $X' = (U', V')$ we have from (3), (8):

$$\begin{aligned} \tilde{d}(Tx, Tx') &= p[\alpha d(U, U') + (1 - \alpha)\, d(V, V')] \\ &\leqq p[\alpha k_1 d(v, v') + \alpha k_2 d(u, u') + (1 - \alpha)\, k_3 d(v, v') + (1 - \alpha)\, k_4 d(u, u')] \\ &= p[\alpha k_2 + (1 - \alpha)\, k_4]\, d(u, u') + p[\alpha k_1 + (1 - \alpha)\, k_3]\, d(v, v') \\ &\leqq pk[\alpha d(u, u') + (1 - \alpha)\, d(v, v')] = k\tilde{d}(x, x') \end{aligned}$$

whenever α satisfies

$$0 < k_4(k - k_2 + k_4)^{-1} \leqq \alpha \leqq (k - k_3)\,(k - k_3 + k_1)^{-1} < 1\,. \tag{9}$$

By force of (H3), there exists at least one such α. QED.

Note that in general $\alpha = 1/2$ does not satisfy (9). A possible alternative choice for $\tilde{d}$ in S is the following:

$$d(x, x') \overset{\text{def}}{=} p \max \{\beta d(u, u'), (1 - \beta)\, d(v, v')\} \tag{3'}$$

then (9) must be replaced by

$$0 < k_4(k - k_3 + k_4)^{-1} \leqq \beta \leqq (k - k_2)\,(k - k_2 + k_1)^{-1} < 1 \tag{9'}$$

under the same assumptions (H1)—(H5), see [9] and Remark 5. We shall henceforth assume (H1)—(H3), (3), (9). Then, denoting by $x_+ = (u_+, v_+)$ the (unique) fixed point of T in $(S, \tilde{d})$: $x_+ = Tx_+$, or $u_+ = T_1(v_+, v_+)\, u_+$, $v_+ = T_2(u_+, u_+)\, v_+$, the usual ("simultaneous") iterative scheme $x_n = Tx_{n-1}$, $x_0 \in S$, $n = 1, 2, \ldots$, or

$$u_n = T_1(v_{n-1}, v_n)\, u_{n-1}\,, \qquad v_n = T_2(u_{n-1}, u_n)\, v_{n-1}\,, \qquad n = 1, 2, \ldots, \tag{A}$$

converges to x_+ for any $x_0 = (u_0, v_0) \in S$. Since $\lambda_2\lambda_5 = 0$ because of (H2), scheme (A) is explicit and can be written in two alternative forms corresponding to $\lambda_2 = 0$ or $\lambda_5 = 0$.

Remark 1: Assumptions (H2), (H3) are invariant with respect to the substitution $u \leftrightarrow v$, which involves the cyclic permutation $\lambda_i \to \lambda_{i+3}$ ($\lambda_{i+6} = \lambda_i$) in (H1). The (dual) iterative schemes obtained by this substitution will be marked in the sequel by a prime.

3. Following [1], let us consider the map $T_2(u, u): S_2 \to S_2$ for an arbitrarily fixed $u \in S_1$. Then for every $v, v' \in S_2$ we have $d(T_2(u, u) v, T_2(u, u) v') \leqq \lambda_6 d(v, v')$ because of (H1), and $\lambda_6 \leqq k_3 < k < 1$ because of (8), (H3). Thus, $T_2(u, u)$ is a contraction in $(S_2, d) = (S_2, d_2)$ for every $u \in S_1$, and its unique fixed point $\bar{v} = T_2(u, u) \bar{v} = Wu$ satisfies $Wu = T_2(u, u) Wu$. It is immediately seen that the operator $W: S_1 \to S_2$ is *Lipschitz continuous*: for every $u, u' \in S_1$

$$d(Wu, Wu') = d(T_2(u, u) Wu, T_2(u', u') Wu') \leqq (\lambda_4 + \lambda_5)(1 - \lambda_6)^{-1} d(u, u'), \tag{10}$$

that is, $\bar{v}$ depends continuously on u.

We now add to (H1)—(H3) the following assumption (H4):

(H4) $\qquad 0 < \bar{k} < k,$

where $\bar{k}$ is defined by

$$\bar{k} = \lambda_3 + (\lambda_1 + \lambda_2)(\lambda_4 + \lambda_5)(1 - \lambda_6)^{-1} \tag{11}$$

and we consider the operator $u \to T_1(Wu, Wu) u$, $u \in S_1$.

Lemma 2:

(i) *Under assumptions* (H1)—(H4), *the operator* $T_1(Wu, Wu): S_1 \to S_1$, $S_1 = (S_1, d_1)$, *is a contraction with constant* $\bar{k} < k$, *and*

(ii) *its unique fixed point* $\bar{u}$ *coincides with* u_+, *while* $\bar{v} = W\bar{u}$ *coincides with* v_+.

Proof:

(i) For every $u, u' \in S_1$ we find from (H1), (10), (11):

$$d(T_1(Wu, Wu) u, T_1(Wu', Wu') u') \leqq (\lambda_1 + \lambda_2) d(Wu, Wu') + \lambda_3 d(u, u')$$
$$\leqq [(\lambda_1 + \lambda_2)(\lambda_4 + \lambda_5)(1 - \lambda_6)^{-1} + \lambda_3] d(u, u') = \bar{k} d(u, u').$$

(ii) It is immediately seen that $\bar{x} = T\bar{x}$, $\bar{x} = (\bar{u}, \bar{v})$, $\bar{v} = W\bar{u}$. QED.

Remark 2: It is easy to verify that assumption (H4) is automatically satisfied if $\lambda_5 = 0$; otherwise, a sufficient condition for (H4) to hold is $(k_2 + k_3) < k$ [9].

According to Lemma 2, the ("consecutive", or "concatenated") iterative scheme [1]

$$u_n = T_1(Wu_{n-1}, Wu_{n-1}) u_{n-1}, \qquad n = 1, 2, \ldots, \tag{B}$$

converges to x_+ for every $u_0 \in S$.

Scheme (B) involves evaluation at each step of the fixed point $v_n = Wu_{n-1}$ of $T_2(u_{n-1}, u_{n-1})$: this theoretically requires an infinite number of iterations, and practically a large number in order to reach a good accuracy (see the detailed analysis in [9]). Thus, the contraction map $T_1(Wu, Wu)$, though of great theoretical value (see e.g. [6, 8] and the related Leray-Schauder method of proof for nonlinear elliptic problems), is not suitable for a numerical evaluation of x_+, in the sense that the amount of computer time required by (B) exceeds by a large factor that involved by (A) [9]. However, (A) may not be the fastest possible iteration scheme, as will be shown below.

Remark 3: The same considerations hold for the dual scheme (B′) obtained from (B) by the substitution $u \leftrightarrow v$. According to Remark 1, its contraction constant $\bar{k}'$ will be assumed to satisfy the dual assumption (H4′):

(H4′) $\qquad 0 < \bar{k}' = \lambda_6 + (\lambda_1 + \lambda_2)(\lambda_4 + \lambda_5)(1 - \lambda_3)^{-1} < k$

which will certainly be fulfilled for $\lambda_2 = 0$ or $k_2 + k_3 < k$ (see Remark 2).

4. Let $W_m: S_1 \to S_2$, $v^{(m)} = W_m u$, be the operator which associates to an arbitrary $u \in S_1$ the m-th iteration $v^{(m)}$ of $T_2(u, u)$: $v^{(m)} = (T_2(u, u))^m v^{(0)}$, or

$$v^{(j)} = T_2(u, u) v^{(j-1)}, \qquad v^{(0)} \in S_2, \qquad j = 1, \ldots, m,$$

(to be precise, we should write $W_m(u, v^{(0)})$, as $v^{(m)}$ is uniquely defined once u and $v^{(0)}$ are prescribed).

Lemma 3:

(i) W_m *is Lipschitz continuous with Lipschitz constant*

$$\lambda_m = (\lambda_4 + \lambda_5)(1 + \lambda_6 + \ldots + \lambda_6^{m-1}) < (\lambda_4 + \lambda_5)(1 - \lambda_6)^{-1};$$

(ii) *the operator* $u \to T_1(W_m u, W_m u) u$, $u \in S_1$, *is contractive with contraction constant* $\bar{k}$;

(iii) *denoting by* $u_+^{(m)} = T_1(W_m u_+^{(m)}, W_m u_+^{(m)}) u_+^{(m)}$ *the (unique) fixed point of the operator considered at* (ii), *and by*

$$v_+^{(m)} = [T_2(u_+^{(m)}, u_+^{(m)})]^m v^{(0)} = W_m u_+^{(m)},$$

for $m \to +\infty$ *we have* $u_+^{(m)} \to u_+$, $v_+^{(m)} \to v_+$ *for any* $v^{(0)} \in S_2$.

Proof: The proof may be carried out as in [1] by means of results obtained in [3], or by adapting the technique of [2], p. 294—295, see [9]. We omit the details for the sake of brevity.

According to Lemma 3, the "diagonalized" [1] iterative scheme

$$u_n = T_1(v_n, v_n) u_{n-1}, \quad u_0 \in S_1; \quad v_n = v_n^{(n)} = v_{n+1}^{(0)} = W_n u_{n-1};; \quad n = 1, 2, \ldots,$$
$$v_n^{(m)} = T_2(u_{n-1}, u_{n-1}) v_n^{(m-1)}, \quad v_0 = v_0^{(0)} = v_1^{(0)} \in S_2; \quad m = 1, \ldots, n,$$

can be proved to be convergent to x_+. This scheme, however, is not too suitable for numerical computations, for it involves a more and more increasing number ($m = n$) of iterations in order to evaluate $W_n u_{n-1}$, see the analogous considerations in [1].

5. We denote by $T^{(q)}: S \to S$ the operator defined by

$$U = T_1(V, V)\, u\,, \qquad V = (T_2(u, u))^q v = W_q u\,, \qquad q \in \mathrm{N}\,, \tag{12}$$

and by $\hat{d}$ the following (equivalent) metric in S:

$$\hat{d}(x, x') \overset{\text{def}}{=} kd(u, u') + \lambda_6^q(\lambda_1 + \lambda_2)\, d(v, v')\,. \tag{13}$$

Lemma 4: *Under assumptions* (H1)—(H5) *below,* $T^{(q)}$ *is a contraction with constant* $h_q < k$ *in the space* $(S, \hat{d})$.

Proof: For every $x = (u, v)$, $x' = (u', v') \in S$ and corresponding $X = (U, V)$, $X' = (U', V')$ in S we have from (H1):

$$d(U, U') = d(T_1(V, V)\, u\,,\ \ T_1(V', V')\, u') \leqq (\lambda_1 + \lambda_2)\, d(V, V') + \lambda_3 d(u, u');$$

$$d(V, V') = d([T_2(u, u)]^q\, v\,,\ [T_2(u', u')]^q\, v') \leqq \ldots < (\lambda_4 + \lambda_5)\,(1 - \lambda_6)^{-1}\, d(u, u') + \lambda_6^q d(v, v')\,.$$

Hence

$$d(U, U') < \bar{k} d(u, u') + (\lambda_1 + \lambda_2)\, \lambda_6^q d(v, v')\,,$$

$$d(V, V') < \bar{k}(\lambda_1 + \lambda_2)^{-1}\, d(u, u') + \lambda_6^q d(v, v')\,.$$

This yields

$$\hat{d}(T^{(q)}x,\ T^{(q)}x') < (k + \lambda_6^q)\, \hat{d}(x, x')\,. \tag{14}$$

On account of (H4), the following inequality (assumption (H5))

H(5) $\qquad h_q \equiv \bar{k} + \lambda_6^q < k$

is satisfied for q large enough, and the assertion follows. QED.

We shall call q *admissible* if it satisfies (H5). According to Lemma 4, the ("intermediate") iterative scheme

$$\begin{aligned} &u_n = T_1(v_n, v_n)\, u_{n-1}\,, \qquad u_0 \in S_1; \qquad v_n^{(q)} = v_n = v_{n+1}^{(0)} = W_q u_{n-1};\\ &v_n^{(m)} = T_2(u_{n-1}, u_{n-1})\, v_n^{(m-1)}\,, \qquad v_0 \in S_2; \qquad m = 1, \ldots, q; \qquad n = 1, 2, \ldots\,, \end{aligned} \tag{C}$$

converges to x_+, for any $x_0 \in S$. This scheme is "intermediate" between (A) and (B), in the sense that at each step a fixed finite number q of iterations is involved in order to calculate $W_q u_{n-1}$ [1].

Remark 4: Interchanging $u \leftrightarrow v$ yields a dual (intermediate) scheme (C') in which a fixed number q' of iterations is required at each step with respect to u, and whose contraction constant h'_q, can easily be derived from h_q according to Remark 1: then, under assumption (H4'), the dual assumption (H5')

(H5') $\qquad h'_{q'} = \bar{k}' + \lambda_3^{q'} < k$

certainly holds for large enough ("admissible") values of $q' \in \mathbb{N}$.

6. From (2), (3), (13), (14) we easily find the error estimates for scheme (A):

$$\tilde{d}(x_n, x_+) \leqq k^n(1 - k)^{-1}\, \tilde{d}(x_1, x_0)\,, \qquad n = 1, 2, \ldots\,, \tag{15}$$

and for scheme (C):

$$\hat{d}(x_n, x_+) < h_q^n(1 - h_q)^{-1}\, \hat{d}(x_1, x_0)\,, \qquad n = 1, 2, \ldots\,, \tag{16}$$

whence the error estimate for the (dual) scheme (C') immediately follows:

$$\bar{d}(x_n, x_+) < (h'_{q'})^n\, (1 - h'_{q'})^{-1}\, \bar{d}(x_1, x_0)\,, \qquad n = 1, 2, \ldots\,, \tag{17}$$

where

$$\bar{d}(x, x') = \lambda_3^{q'}(\lambda_4 + \lambda_5)\, d(u, u') + \bar{k}' d(v, v') \tag{18}$$

is the (equivalent) metric for S such that the dual of Lemma 4 holds.

We might now compare the rates of convergence of schemes (A), (C), (C') by means of their root convergence factors as in [1] (see [9] for details). What is of practical interest, however, is not so much the rate of convergence as the amount of computer time. As in [1], let us denote by: a_i the (computer) time required to carry out one iteration with respect to the operator T_i, $i = 1, 2$; $n(\mathrm{J}, \varepsilon)$ the number of iterations required by scheme (J) in order to achieve an approximation to x_+ with an error less than or equal to $\varepsilon > 0$:

$$\tilde{d}(x_n, x_+) \leqq \varepsilon\,, \qquad n = n(\mathrm{J}, \varepsilon)\,, \tag{19}$$

with respect to the metric $\tilde{d}$; $T(\mathrm{J}, \varepsilon)$ the amount of computer time required by J in correspondence to the error (19); finally, for every real y let $[y]$ denote the largest integer $<y$ if $y > 0$, 0 if $y \leqq 0$. We shall give upper bounds for $n(\mathrm{J}, \varepsilon)$ and $T(\mathrm{J}, \varepsilon)$ for (J) = (A), (C), (C').

By a suitable definition of the constant $b_1 > 0$ we find from (15), (19):

$$n(A, \varepsilon) \leqq 1 + [\log\,(b_1\varepsilon)/\log\, k]\,, \tag{20}$$

$$T(\mathrm{A}, \varepsilon) = n(\mathrm{A}, \varepsilon)\,(a_1 + a_2) \leqq \{1 + [\log(b_1\varepsilon)/\log\, k]\}\,(a_1 + a_2) \equiv \tilde{T}(\mathrm{A}, \varepsilon)\,.$$

By a suitable definition of $b_2 > 0$ ($\tilde{d}$ and $\hat{d}$ are equivalent) we find from (16), (19):

$$n(C, \varepsilon) < 1 + [\log (b_2\varepsilon)/\log h_q], \tag{21}$$

$$T(C, \varepsilon) = n(C, \varepsilon)(a_1 + qa_2) < \{1 + [\log (b_2\varepsilon)/\log h_q]\}(a_1 + qa_2) \equiv \tilde{T}(C, \varepsilon),$$

whence for a suitable $b_3 > 0$ ($\tilde{d}$ and $\bar{d}$ are equivalent):

$$T(C', \varepsilon) = n(C', \varepsilon)(q'a_1 + a_2) < \{1 + [\log (b_3\varepsilon)/\log h'_{q'}]\}(q'a_1 + a_2) = \tilde{T}(C', \varepsilon).$$

Here, the values of q and q' are admissible ones, that is, they satisfy (H5) and (H5'), and so $|\log h_q| > |\log k| < < |\log h'_{q'}|$. From (20) and (21) we see that $\tilde{T}(C, \varepsilon) < \tilde{T}(A, \varepsilon)$ ($>$) if

$$a_2(q - 1 + q[\log (b_2\varepsilon)/\log h_q] - [\log (b_1\varepsilon)/\log k]) < a_1([\log (b_1\varepsilon)/\log k] - [\log (b_2\varepsilon)/\log h_q]) \; (>). \tag{22}$$

Thus, (22) prescribes in general an upper bound for a_2/a_1 in order that $\tilde{T}(C, \varepsilon) < \tilde{T}(A, \varepsilon)$. In the asymptotic case [1] $\varepsilon \to 0^+$ (22) becomes, omitting integer parts:

$$(a_2/a_1)(q|\log k| - |\log h_q|) < |\log h_q| - |\log k|,$$

which is always satisfied if $h_q < k^q$. The following theorem yields additional informations about "asymptotically optimal" [1] schemes.

Theorem 1: *Let* (H1)–(H4), (H4') *be satisfied, and* $\varepsilon \to 0^+$.

(i) *If* $a_1 = a_2$, $k^{1+q} < h_q^2$, $k^{1+q'} < (h'_{q'})^2$ *for all admissible* q, q' ($q > 1$, $q' > 1$) *then* (A) *has a lower upper bound for* $T(\varepsilon)$ *than* (C) *and* (C') *for all admissible* q *and* q'.

(ii) *If* a_2/a_1 *is small enough, then there exists an admissible value* q_0 *such that* $q_0a_2 \ll a_1$ *and* $\tilde{T}(C, \varepsilon)$ *has a minimum at* $q = q_0$ *as* q *varies* (q *admissible*). *Then if* $(h_{q_0})^{q'} < h'_{q'}$ *for all admissible* q', *scheme* (C) *with* $q = q_0$ *has the lowest upper bound for* $T(\varepsilon)$ *among schemes of type* (C), (C'), (A).

(iii) *If* a_1/a_2 *is small enough, a dual result for* (C') *holds.*

We omit the proof, since it is tedious but straightforward.

A numerical example, wherein $S = \mathbb{R}^2$ and the assumptions of Theorem 1, (ii) are satisfied, is reported in [9].

Remark 5: Assume S_1, S_2 are Hilbert spaces, the operators T_1, T_2 are Lipschitz continuous (as in (H1)) and strongly monotone [16], and let the norm in S be defined according to (3'). Then the operator U_t: $S \to S$, $U_tx = x - tTx$, $t > 0$, is a contraction with constant $K(t) < 1$ for a certain interval of values of t, $0 < t < t^*$, and $K(t)$ is minimal under certain conditions [16].

Acknowledgment

This research was supported by the GNFM of CNR.

References

1 SCHNEIDER, K. R. Zur Konvergenzbeschleunigung bei Summen strikt kontraktiver Operatoren, ZAMM 58, 383–389 (1978).
2 LANGENBACH, A., Monotone Potentialoperatoren, Springer-Verlag Berlin-New York 1977
3 GAJEWSKI, H., Iterationsverfahren bei Folgen strikt kontraktiver Operatoren, Monatsber. Deutsch. Akad. Wiss. 11, 818–826 (1969).
4 KLUGE, R., Zur approximativen Lösung nichtlinearer Variationsgleichungen, Dissertation, Humboldt Universität, Berlin 1970.
5 CINQUINI-CIBRARIO, M.; CINQUINI, S.; Equazioni a derivate parziali di tipo iperbolico, Cremonese, Roma 1964.
6 CESARI, L., A boundary value problem for quasilinear hyperbolic systems in the Schauder canonic form. Ann Sci Norm Sup. Pisa, IV. Ser 1, 311–358 (1974).
7 BASSANINI, P., Iterative methods for quasilinear hyperbolic systems, to be published on Boll. U M.I.
8 BASSANINI, P.; CESARI, L., La duplicazione di frequenza nella radiazione laser, to be published on Atti Accad. Naz. Lincei.
9 BASSANINI, P.; FILIAGGI, E., Schemi iterativi e accelerazione della convergenza per operatori di contrazione nel prodotto di due spazi di BANACH, Atti Sem. Mat. Fis. Univ. Modena 28, 249–279 (1979)
10 SANCHEZ, D. A., An iteration scheme for HILBERT space boundary value problems, Boll. U M.I., IV Ser. 11, 1–9 (1975).
11 CESARI, L., Alternative methods in nonlinear analysis, in: ANTOSIEWICZ, H. A. (Ed.), Internat. Conf. on Differential Equations. Academic Press, New York 1975, p. 95–148.
12 CESARI, L., Nonlinear oscillations under hyperbolic systems, in: CESARI, HALE, LASALLE (Eds.), Internat. Conf. on Dynamical Systems, Academic Press, New York 1976, Vol 1, 251–261
13 MAWHIN, J., Boundary value problems at resonance for vector second order nonlinear ordinary differential equations, in: EQUADIFF IV, Lecture Notes in Mathematics ≠ 703, Springer, Berlin 1979, 241–249.
14 BANFI, C.; CASADEI, G., Calcolo di soluzioni periodiche di equazioni differenziali non lineari, Calcolo 5, Suppl. 1, 1–11 (1968).
15 WILLIAMS, S. A., A sharp sufficient condition for solutions of a nonlinear elliptic boundary value problem, J. Differential Equations 8, 580–586 (1970).
16 GAJEWSKI, H.; GRÖGER, K., Zur iterativen Lösung von Gleichungen mit stark monotonen LIPSCHITZ-stetigen Operatoren, Math. Nachr 67, 273–280 (1975).

Eingereicht am 2. 1 1981

Anschrift Prof Dr PIERO BASSANINI, Istituto Matematico, Università di Roma, P le A Moro 5, 00185 ROMA, Italy

ZAMM 62, 7 – 13 (1982)

N. Schneider

Monotone Enclosure Using Multi-Point-Secant Methods

Um obere bzw. untere Schrankenfolgen für die Lösung von Gleichungssystemen zu erhalten, werden Mehrschrittverfahren betrachtet. Es wird gezeigt, daß diese Folgen monoton und überlinear konvergieren. Die erzielten Ergebnisse werden an Hand eines Beispieles demonstriert.

Multi-point-secant methods are applied to systems of equations in order to obtain sequences of upper and lower bounds for the solution. It is shown that the sequences converge monotonically and the order of convergence is better than linear. The results are illustrated by an example.

Рассматриваются многошаговые процедуры чтобы получить последовательности верхних или нижних граней для решения систем уравнений. Доказывается, что сходимость этих последовательностей является монотоной и суперлинейной. Полученные результаты иллюстрироваются с помощью примера.

1. Introduction

For the iterative solution of n-dimensional systems of nonlinear equations there are considered in [7] and [10] multi-point-secant methods. These methods work with approximations of the Jacobian matrix, which differ in consecutive iteration steps only by a matrix of a fixed rank l, $1 \leqq l \leqq n$. The effectivity of these methods is higher than the effectivity of standard methods as for instance the regula falsi, if the number of function value computations is considered.

In [7] and [10] there are proved local convergence results for these methods. In this paper we first generalize monotone enclosure statements, which are proved in [2], [8], [9]. This generalization is also used to prove monotone enclosure results for a multi-point-secant method, which is developed from one of the mentioned methods in [7], [10].

2. Preliminaries

The definitions and properties used in connection with the cone, which introduces a partial ordering in a Banach space B, are found in [11]. The i-th component of a vector x is denoted by $x^{(i)}$. The j-th column of an $n \times n$ matrix A is denoted by $A^{(j)}$. For $x, y \in \mathbb{R}^n$, $x \leqq y$ means $x^{(i)} \leqq y^{(i)}$, $i = 1(1)\,n$. The notions "comparable", "isoton" can be found e.g. in [3]. $S(B)$ means the set of all continuous, linear operators $(B \to B)$.

Kantorovich's *Lemma* [1]: *Let B be a Banach space, which is partially ordered by a closed, regular cone and $A([x, y] \subset B \to B)$ a continuous, isotone mapping. If $x \leqq A(x)$ and $A(y) \leqq y$, then A has a fixed point in $[x, y]$.*

3. General enclosure statements

In what follows B denotes a Banach space, which is partially ordered by a closed, regular cone. Schmidt [6] defines a first divided difference operator of $F(D \subset B \to B)$ as a mapping $\delta F(D \times D \subset B \times B \to S(B))$, which satisfies

$$\delta F(x, y)\,(x - y) = F(x) - F(y) \quad \text{for all } x, y\,.$$

We will now generalize Schmidt's concept of a divided difference operator in the following manner.

Definition 3.1: A mapping $A(D \times D \subset B \times B \to S(B))$ is called *generalized divided difference operator* („verallgemeinerte Steigung") of $F(D \subset B \to B)$ if

$$A(x, y)\,(x - y) \geqq F(x) - F(y) \quad \text{for all comparable } x, y\,. \tag{3.1}$$

Using this generalization the following theorem was proved in [9]. It generalizes an enclosure theorem proved in [8].

Theorem 3.1: *Let F be a continuous mapping $(V := [x_0, y_1] \subset B \to B)$. Suppose there is an $x_1 \in V$ such that $F(x_1) \geqq 0 \geqq F(y_1)$. We consider the iteration*

$$\left.\begin{aligned} F(x_k) + A(x_k, x_{k-1})\,(x_{k+1} - x_k) = 0\,,\\ F(y_k) + A(x_{k+1}, x_k)\,(y_{k+1} - y_k) = 0\,,\end{aligned}\right\} \quad k = 1, 2, \ldots, \tag{3.2}$$

and assume that $A(V \times V \to S(B))$ is a generalized divided difference operator of F with the property

$$A(u_1, v_1) \geqq A(u_2, v_2) \quad \text{for} \quad u_1 \geqq u_2\,, \qquad v_1 \geqq v_2\,. \tag{3.3}$$

It there exists a mapping $T \in S(B)$, which is nonnegative and injective, and an operator $G(B \to B)$ such that

$$T(G + A(x_1, x_0)) \geqq 0\,, \qquad TG \leqq I\,, \tag{3.4}$$

then the iteration (3.2) *is well defined, that means there exist solutions x_{k+1}, y_{k+1}. The monotone sequences (x_k), (y_k) have limits x^*, y^*, which are solutions of $F(x) = 0$, and the monotone enclosing*

$$x_1 \leqq \cdots \leqq x_k \leqq x_{k+1} \leqq \cdots \leqq x^* \leqq z^* \leqq y^* \leqq \cdots \leqq y_{k+1} \leqq y_k \leqq \cdots \leqq y_1 \tag{3.5}$$

holds for any solution $z^* \in [x_1, y_1]$ of $F(x) = 0$.

Using this theorem it was possible for the first time to give conditions under which the sequences (x_k), (y_k) generated by

$$\left.\begin{aligned} F(x_k) + \overline{\delta F}(x_k, x_{k-1})(x_{k+1} - x_k) = 0,\\ F(y_k) + \overline{\delta F}(x_{k+1}, x_k)(y_{k+1} - y_k) = 0, \end{aligned}\right\} \quad k=1, 2, \dots, \tag{3.6}$$

— iteration (3.2) with $A = \overline{\delta F}$ — enclosure monotonically the solution of $F(x) = 0$, $x \in \mathbb{R}^n$. Here $\overline{\delta F}(u, v)$ is an $n \times n$ matrix — a well known and frequently used approximation of the Jacobian matrix — with the columns defined by

$$\overline{\delta F}^{(j)}(u, v) := \begin{cases} \dfrac{F(u) - F(u - (u^{(j)} - v^{(j)})e^{(j)})}{u^{(j)} - v^{(j)}}, & u^{(j)} \neq v^{(j)},\\[2ex] \dfrac{\partial F}{\partial x^{(j)}}(u), & u^{(j)} = v^{(j)}. \end{cases} \tag{3.7}$$

$e^{(j)}$, $j = 1(1)n$ denote the unit coordinate vectors of $\mathbb{R}^n$ and $\dfrac{\partial F(x)}{\partial x^{(j)}}$ the j-th column of the Jacobian matrix $F'(x)$.

We consider now the iteration method

$$\left.\begin{aligned} F(x_k) + A_k(x_k, x_{k-1})(x_{k+1} - x_k) = 0,\\ F(y_k) + A_{k+1}(x_{k+1}, x_k)(y_{k+1} - y_k) = 0, \end{aligned}\right\} \quad k = 1, 2, \dots, \tag{3.8}$$

which is a generalization of method (3.2).

In the following theorem we will prove for method (3.8) monotone enclosure statements, which generalize the results of Theorem 3.1. A realization of iteration method (3.8) was given in [2] by using interval analysis. The enclosure statements, which are proved in [2], are a special case of the results of Theorem 3.2.

Theorem 3.2: *Let $F(V := [x_0, y_1] \subset B \to B)$ be a continuous mapping. Suppose there is a $x_1 \in V$ such that $F(x_1) \geqq 0 \geqq F(y_1)$ and a solution $z^* \in [x_1, y_1]$ of $F(x) = 0$. We consider method (3.8) with operators*

$$A_k(M_k := \{(u, v) \mid x_{k-1} \leqq v \leqq u \leqq y_1\} \to S(B))$$

such that

$$A_k(u, v) \leqq A(u, v), \tag{3.9}$$

where A is a generalized divided difference operator of F with the property

$$A(u_1, v_1) \geqq A(u_2, v_2) \quad \text{for} \quad u_1 \geqq u_2, \quad v_1 \geqq v_2. \tag{3.10}$$

If there are nonnegative, injective mappings $T_k \in S(B)$ and mappings $G_k(B \to B)$ such that

$$T_k\{G_k + A_k(u, v)\} \geqq 0, \qquad T_k G_k \leqq I, \tag{3.11}$$

then the iteration (3.8) is well-defined. That means, that solutions x_{k+1}, y_{k+1} of the linear equations (3.8) exist.

The monotone sequences (x_k), (y_k) have limits x^*, y^*, and if $\lim T_k$ exists and is injective too, they are solutions of $F(x) = 0$.

The monotone enclosing

$$x_1 \leqq \dots \leqq x_k \leqq x_{k+1} \leqq \dots \leqq x^* \leqq z^* \leqq y^* \leqq \dots \leqq y_{k+1} \leqq y_k \leqq \dots \leqq y_1 \tag{3.12}$$

holds for any solution $z^* \in [x_1, y_1]$ of $F(x) = 0$.

Proof: It will be proved by induction that

$$x_{k-1} \leqq x_k \leqq z^* \leqq y_k, \qquad F(x_k) \geqq 0 \geqq F(y_k), \tag{3.13}$$

where z^* is a solution of $F(x) = 0$ in $[x_1, y_1]$, which is assumed to exist.

(3.13) is correct for $k = 1$. For proving (3.13) for all k, we consider the continuous operator H_k defined by

$$H_k(z) := z + T_k\{F(x_k) + A_k(x_k, x_{k-1})(z - x_k)\}, \qquad z \in [x_k, z^*].$$

$z_1 \geqq z_2$ implies

$$H_k(z_1) - H_k(z_2) = z_1 - z_2 + T_k A_k(x_k, x_{k-1})(z_1 - z_2) \geqq T_k\{G_k + A_k(x_k, x_{k-1})\}(z_1 - z_2) \geqq 0,$$

so H_k is also isotone. Together with

$$H_k(x_k) = x_k + T_k F(x_k) \geqq x_k,$$

$$\begin{aligned} H_k(z^*) &= z^* + T_k\{F(x_k) + A_k(x_k, x_{k-1})(z^* - x_k)\} \leqq z^* + T_k\{F(x_k) + A(x_k, x_{k-1})(z^* - x_k)\}\\ &\leqq z^* + T_k\{F(x_k) + A(x_k, z^*)(z^* - x_k)\} \leqq z^* + T_k F(z^*) = z^* \end{aligned}$$

the Kantorovich Lemma guarantees the existence of a fixed point $x_{k+1} \in [x_k, z^*]$, which is a solution of the equation

$$F(x_k) + A_k(x_k, x_{k-1})(x_{k+1} - x_k) = 0.$$

Therefore we obtain

$$\begin{aligned} F(x_{k+1}) &\geqq A(x_k, x_{k+1})(x_{k+1} - x_k) + F(x_k) \geqq A(x_k, x_{k-1})(x_{k+1} - x_k) + F(x_k)\\ &\geqq A_k(x_k, x_{k-1})(x_{k+1} - x_k) + F(x_k) = 0. \end{aligned}$$

We now define the continuous and isotone operator $\overline{H}_{k+1}$

$$\overline{H}_{k+1}(z) := z + T_{k+1}\{F(y_k) + A_{k+1}(x_{k+1}, x_k)(z - y_k)\}, \qquad z \in [z^*, y_k].$$

Because of

$$\overline{H}_{k+1}(z^*) = z^* + T_{k+1}\{F(y_k) + A_{k+1}(x_{k+1}, x_k)(z^* - y_k)\} \geqq z^* + T_{k+1}\{F(y_k) + A(x_{k+1}, x_k)(z^* - y_k)\}$$
$$\geqq z^* + T_{k+1}\{F(y_k) + A(z^*, y_k)(z^* - y_k)\} \geqq z^* + T_{k+1}F(z^*) = z^*,$$
$$\overline{H}_{k+1}(y_k) = y_k + T_{k+1}F(y_k) \leqq y_k,$$

the Kantorowich Lemma guarantees the existence of a fixed point $y_{k+1} \in [z^*, y_k]$, which is a solution of the equation

$$F(y_k) + A_{k+1}(x_{k+1}, x_k)(y_{k+1} - y_k) = 0$$

and has the property

$$F(y) \leqq A(y_{k+1}, y_k)(y_{k+1} - y_k) + F(y_k) \leqq A(x_{k+1}, x_k)(y_{k+1} - y_k) + F(y_k)$$
$$\leqq A_{k+1}(x_{k+1}, x_k)(y_{k+1} - y_k) + F(y_k) = 0.$$

The statement (3.13) is proved and it has been shown that the sequences (x_k), (y_k) are monotone and enclose all solutions of $F(x) = 0$ in $[x_1, y_1]$.

By the regularity of the cone the limits $x^* = \lim x_k$, $y^* = \lim y_k$ exist. We have

$$0 \leqq T_k \cdot F(x_k) = -T_k A_k(x_k, x_{k-1})(x_{k+1} - x_k) \leqq T_k G_k(x_{k+1} - x_k) \leqq x_{k+1} - x_k,$$
$$0 \geqq T_{k+1} \cdot F(y_k) = -T_{k+1} A_{k+1}(x_{k+1}, x_k)(y_{k+1} - y_k) \geqq T_{k+1} G_{k+1}(y_{k+1} - y_k) \geqq y_{k+1} - y_k$$

and therefore we obtain by the continuity of the operators F and T_k

$$TF(x^*) = TF(y^*) = 0.$$

Using the injectivity of T it follows $F(x^*) = F(y^*) = 0$.

Remark: By setting $A_k = A$ we obtain Theorem 3.1.

There are other versions of Theorem 3.2 corresponding to different sign configurations. We indicate these versions schematically in Table 3.1, where the first row represents Theorem 3.2.

We define: $(i = 0, 1)$

$$A_i: \begin{cases} (-1)^i x_0 \leqq (-1)^i x_1 \leqq (-1)^i y_1, \\ M_k := \{(u, v) \mid (-1)^i x_{k-1} \leqq (-1)^i v \leqq (-1)^i u \leqq (-1)^i y_1\}; \end{cases}$$

$$B_i: \begin{cases} (-1)^i F(x_1) \geqq 0 \geqq (-1)^i F(y_1), \\ (-1)^i A(u_1, v_1) \geqq (-1)^i A(u_2, v_2), \qquad u_1 \geqq u_2, \qquad v_1 \geqq v_2; \end{cases}$$

$$C_i: \quad \{(-1)^i A_k(u, v) \leqq (-1)^i A(u, v);$$

$$D_i: \begin{cases} \text{There are mappings } T_k \in S(B) \text{, which are nonnegative, injective left subinverses of operators} \\ G_k^i(B \to B) \text{ such that} \\ T_k\{G_k^i + (-1)^i A_k(u, v)\} \geqq 0; \end{cases}$$

$$E_i: \quad \{(-1)^i x_k \leqq (-1)^i x_{k+1} \leqq (-1)^i z^* \leqq (-1)^i y_{k+1} \leqq (-1)^i y_k.$$

Table 3.1

I	A_0,	B_0,	C_0,	$D_0 \Rightarrow E_0$
II	A_1,	B_0,	C_1,	$D_1 \Rightarrow E_1$
III	A_0,	B_1,	C_1,	$D_1 \Rightarrow E_0$
IV	A_1,	B_1,	C_0,	$D_0 \Rightarrow E_1$

Remark: If we replace the corresponding assumptions of Theorem 3.2 as indicated in Table 3.1, the enclosure statements E_i follow.

In the following chapter we will show by using the results of Theorem 3.2 that it is possible to prove monotone enclosure statements for the multi-point-secant methods, which are described in [7].

4. Monotone enclosure using multi-point-secant methods

One of the multi-point-secant methods, which are described in [7] and [10] is the $(n + 1)$-*point-secant method*

$$F(x_k) + B_k(x_k, x_{k-1})(x_{k+1} - x_k) = 0, \qquad k = 1, 2, \ldots$$

$B_k(u, v)$ denotes an $n \times n$ matrix with the columns defined by

$$l \in \{3, \ldots, n\}: B_k^{(j)}(u, v) := \overline{\delta F^{(j)}}(x_{k-l+1}, x_{k-l}),$$
$$l = 2: B_k^{(j)}(u, v) := \overline{\delta F^{(j)}}(v, x_{k-l}),$$
$$l = 1: B_k^{(j)}(u, v) := \overline{\delta F^{(j)}}(u, v),$$
$$j := k - l - 1 \pmod{n} + 1, \qquad l \in \{1, \ldots, n\}.$$

For the definition of the $n \times n$ matrix $\overline{\delta F}(u, v)$ see (3.7).

It is obvious that for constructing $B_k(u, v)$, we have to replace only one column by crossing from $B_k(x_k, x_{k-1})$ to $B_{k+1}(x_{k+1}, x_k)$ and due to the applicability of the Sherman-Morrison formula [7], the solution of the linear equations required in each step of the iteration can be performed with less effort than by using Gaussian elimination.

We consider now the iteration method

$$\left.\begin{aligned} F(x_k) + B_k(x_k, x_{k-1})(x_{k+1} - x_k) = 0, \\ F(y_k) + B_{k+1}(x_{k+1}, x_k)(y_{k+1} - y_k) = 0, \end{aligned}\right\} \quad k = 1, 2, \dots \tag{4.1}$$

Iteration method (4.1) is a realization of method (3.8) and by using the results of Theorem 3.2 we will now prove monotone enclosure statements for iteration method (4.1).

Theorem 4.1: *Let $F(V := [x_{-n}, y_1] \subset \mathbb{R}^n \to \mathbb{R}^n)$ be a continuous mapping. Suppose that the operator $\overline{\delta F}(V \times V \to S(\mathbb{R}^n))$ is a generalized divided difference operator of F and has the property*

$$\overline{\delta F}(u_1, v_1) \geqq \overline{\delta F}(u_2, v_2) \quad \text{for} \quad u_1 \geqq u_2, \quad v_1 \geqq v_2. \tag{4.2}$$

In addition, assume there are $x_l \in V(l = -n + 1(1) - 1)$, x_0, x_1 such that

$$x_l \leqq x_0 \leqq x_1, \qquad F(x_1) \geqq 0 \geqq F(y_1).$$

If there exists a mapping $T \in S(\mathbb{R}^n)$, which is nonnegative and injective, and an operator $G(\mathbb{R}^n \to \mathbb{R}^n)$ such that

$$T\{G + \overline{\delta F}(x_{-n}, x_{-n})\} \geqq 0, \qquad TG \leqq I,$$

then the iteration (4.1) is well-defined, the monotone sequences (x_k), (y_k) have limits x^*, y^*, which are solutions of $F(x) = 0$ and the monotone enclosing

$$x_1 \leqq \dots \leqq x_k \leqq x_{k+1} \leqq \dots \leqq x^* \leqq z^* \leqq y^* \leqq \dots \leqq y_{k+1} \leqq y_k \leqq \dots \leqq y_1, \tag{4.3}$$

holds for any solution $z^* \in [x_1, y_1]$ of $F(x) = 0$.

Proof: We show that the conditions of Theorem 3.2 are realized with $A := \overline{\delta F}$, $A_k := B_k$, $G_k := G$, $T_k := T$.

First we prove that there exists a solution of $F(x) = 0$, $x \in [x_1, y_1]$. We consider the continuous operator H defined by

$$H(z) := z + TF(z), \qquad z \in [x_1, y_1].$$

$z_1 \geqq z_2$ implies

$$\begin{aligned} H(z_1) - H(z_2) &= z_1 - z_2 + T\{F(z_1) - F(z_2)\} \geqq z_1 - z_2 + T\overline{\delta F}(z_2, z_1)(z_1 - z_2) \\ &\geqq T\{G + \overline{\delta F}(z_2, z_1)\}(z_1 - z_2) \geqq T\{G + \overline{\delta F}(x_{-n}, x_{-n})\}(z_1 - z_2) \geqq 0, \end{aligned}$$

therefore H is isotone. Together with

$$H(x_1) = x_1 + TF(x_1) \geqq x_1, \qquad H(y_1) = y_1 + TF(y_1) \leqq y_1$$

the Kantorovich Lemma guarantees the existence of a fixed point $z^* \in [x_1, y_1]$, which is a solution of $F(x) = 0$.

We consider $B_1(M_1 := \{(u, v) \mid x_0 \leqq v \leqq u \leqq y_1\} \to S(\mathbb{R}^n))$ and get $B_1(u, v) \leqq \overline{\delta F}(u, v)$, because of

$$\begin{aligned} &B_1^{(j)}(u, v) = \overline{\delta F}^{(j)}(x_{-n+j+1}, x_{-n+j}) \leqq \overline{\delta F}^{(j)}(u, v), \qquad j \in \{1, \dots, n-2\}, \\ &B_1^{(n-1)}(u, v) = \overline{\delta F}^{(n-1)}(v, x_0) \leqq \overline{\delta F}^{(n-1)}(u, v), \\ &B_1^{(n)}(u, v) = \overline{\delta F}^{(n)}(u, v). \end{aligned}$$

We also get by (4.2)

$$B_1(u, v) \geqq \delta F(x_{-n}, x_{-n})$$

and therefore

$$T\{G + B_1(u, v)\} \geqq T\{G + \overline{\delta F}(x_{-n}, x_{-n})\} \geqq 0.$$

We show by induction

$$\left.\begin{aligned} &T\{G + B_k(u, v)\} \geqq 0, \\ &B_k(u, v) \leqq \overline{\delta F}(u, v), \end{aligned}\right. \qquad (u, v) \in M_k := \{(u, v) \mid x_{k-1} \leqq v \leqq u \leqq y_1\}, \tag{4.4}$$

which is correct for $k = 1$.

If the statement (4.4) is correct for all $k < \bar{k}$, all conditions of Theorem 3.2 are realized for all $k < \bar{k}$ and therefore we get $x_k \leqq x_{k+1}$ for all $k < \bar{k}$ (see the proof of Theorem 3.2). It follows for $(u, v) \in M_{\bar{k}}$

$$\begin{aligned} &T\{G + B_{\bar{k}}(u, v)\} \geqq T\{G + \overline{\delta F}(x_{-n}, x_{-n})\} \geqq 0, \\ &B_{\bar{k}}^{(j)}(u, v) = \overline{\delta F}^{(j)}(x_{\bar{k}-l+1}, x_{\bar{k}-l}) \leqq \overline{\delta F}^{(j)}(u, v), \qquad l \in \{3, \dots, n\}, \\ &B_{\bar{k}}^{(j)}(u, v) = \overline{\delta F}^{(j)}(v, x_{\bar{k}-l}) \leqq \overline{\delta F}^{(j)}(u, v), \qquad l = 2, \\ &B_{\bar{k}}^{(j)}(u, v) = \overline{\delta F}^{(j)}(u, v), \qquad l = 1. \end{aligned}$$

Therefore the conditions of Theorem 3.2 are realized for all k, and the statements are proved by Theorem 3.2.

Remarks: $\overline{\delta F}$ is a generalized divided difference operator with property (4.2) if the second partial derivatives of F are nonnegative [9]. If there are different sign configurations, one can prove the other versions of Theorem 4.1 using Table 3.1. We see that the same assumptions as for iteration method (3.6) guarantee the enclosure statements for method (4.1).

It is obvious that the same results are obtained, if we consider the iteration

$$\left.\begin{aligned} &F(x_k) + B_k(x_k, x_{k-1})\,(x_{k+1} - x_k) = 0\,,\\ &F(y_k) + B_{k+1}(x_{k+1}, x_k)\,(y_{k+1} - y_k) = 0\,,\end{aligned}\right\}\quad k = 1, 2, \ldots, \tag{4.5}$$

using an $n \times n$ matrix with the columns defined by

$$\left.\begin{aligned} &l \in \{3, \ldots, n\}\colon B_k^{(j)}(u, v) := \delta F^{(j)}(x_{k-l+1}, x_{k-l})\,,\\ &l = 2\colon B_k^{(j)}(u, v) := \delta F^{(j)}(v, x_{k-l})\,,\\ &l = 1\colon B_k^{(j)}(u, v) := \delta F^{(j)}(u, v)\,,\\ &j := k - l - 1 \pmod n + 1\,,\qquad l \in \{1, \ldots, n\}\,.\end{aligned}\right|$$

$\delta F(u, v)$ means the realisation of SCHMIDT's concept of a divided difference operator. It is an $n \times n$ matrix with the columns defined by

$$\delta F^{(j)}(u, v) = \begin{cases} \dfrac{F(\ldots, v^{(j-1)}, u^{(j)}, u^{(j+1)}, \ldots) - F(\ldots, v^{(j-1)}, v^{(j)}, u^{(j+1)}, \ldots)}{u^{(j)} - v^{(j)}}\,, & u^{(j)} \neq v^{(j)}\\[2ex] \dfrac{\partial F}{\partial x^{(j)}}\,(v^{(1)}, \ldots, v^{(j-1)}, u^{(j)}, u^{(j+1)}, \ldots, u^{(n)})\,, & u^{(j)} = v^{(j)}\,.\end{cases}$$

Notice that the effectivity of iteration (4.1) is higher than the effectivity of (4.5), because (4.5) needs one function evaluation more per iteration step.

Further methods can be obtained in the same manner like (4.1) from the other methods, which are described in [7] and the results of this chapter hold for those methods analogously.

5. Rate-of-convergence and computational efficiency

In this chapter we will show a statement about the convergence order of iteration (4.1) and compare the computational efficiency of (4.1) with those of the 2-point-method (3.6).

Theorem 5.1: *Let the assumptions of Theorem* 4.11 *be valid. If we assume that there exists a mapping* $S(\mathbb{R}^n \to \mathbb{R}^n)$ *with the properties*

$$S \leqq -\overline{\delta F}(u, v) \quad \text{for} \quad u \geqq v \quad \text{and} \quad S^{-1} \geqq 0\,, \tag{5.1}$$

then

$$x^* = \lim x_k = \lim y_k = y^*\,.$$

If, in addition, $S^{-1} \in S(\mathbb{R}^n)$ *and* $\overline{\delta F}$ *satisfies*

$$\|\overline{\delta F}(u, v) - \overline{\delta F}(v, w)\| \leqq \beta(\|u - v\| + \|u - w\| + \|v - w\|) \quad \text{for} \quad u \geqq v \geqq w\,, \tag{5.2}$$

then the R-order of iteration (4.1) *at* x^* *is not less than* τ, *where* τ *is the unique positive root of* $t^{n+1} - t^n - 1 = 0$.

Proof: We know from Theorem 4.1 that $x^* = \lim x_k \leqq \lim y_k = y^*$. From

$$0 = F(y^*) - F(x^*) \leqq \overline{\delta F}(y^*, x^*)\,(y^* - x^*) \leqq -S(y^* - x^*)$$

it follows

$$0 \leqq S^{-1}\{-S(y^* - x^*)\} = x^* - y^*\,,$$

and therefore

$$x^* = y^*\,.$$

For $k \geqq n$ from

$$\begin{aligned} S(x^* - x_{k+1}) &\leqq -\overline{\delta F}(x_{k-n+1}, x_{k-n})\,(x^* - x_{k+1})\\ &= \overline{\delta F}(x_{k-n+1}, x_{k-n})\,(x_{k+1} - x_k) - \overline{\delta F}(x_{k-n+1}, x_{k-n})\,(x^* - x_k)\\ &\leqq B_k(x_k, x_{k-1})\,(x_{k+1} - x_k) - \overline{\delta F}(x_{k-n+1}, x_{k-n})\,(x^* - x_k)\\ &= F(x^*) - F(x_k) - \overline{\delta F}(x_{k-n+1}, x_{k-n})\,(x^* - x_k)\\ &\leqq \{\overline{\delta F}(x^*, x_k) - \overline{\delta F}(x_{k-n+1}, x_{k-n})\}\,(x^* - x_k)\\ &= \{\overline{\delta F}(x^*, x_k) - \overline{\delta F}(x_k, x_{k-n+1}) + \overline{\delta F}(x_k, x_{k-n+1}) - \delta F(x_{k-n+1}, x_{k-n})\}\,(x^* - x_k)\end{aligned}$$

it follows that

$$0 \leqq x^* - x_{k+1} \leqq S^{-1}\{\overline{\delta F}(x^*, x_k) - \overline{\delta F}(x_k, x_{k-n+1}) + \overline{\delta F}(x_k, x_{k-n+1}) - \overline{\delta F}(x_{k-n+1}, x_{k-n})\}\,(x^* - x_k)\,.$$

From this we get by the continuity of S^{-1} and (5.2) together with the triangle inequality

$$\|x^* - x_{k+1}\| \leqq \gamma \cdot \|x^* - x_k\| \cdot \|x^* - x_{k-n}\| \ . \tag{5.3}$$

It follows in a similar manner that

$$\begin{aligned} S(y_{k+1} - x^*) &\leqq -\delta\overline{F}(x_{k-n+1}, x_{k-n})\,(y_{k+1} - x^*) \\ &= -\delta\overline{F}(x_{k-n+1}, x_{k-n})\,(y_{k+1} - y_k) - \delta\overline{F}(x_{k-n+1}, x_{k-n})\,(y_k - x^*) \\ &\leqq -B_{k+1}(x_{k+1}, x_k)\,(y_{k+1} - y_k) - \delta\overline{F}(x_{k-n+1}, x_{k-n})\,(y_k - x^*) \\ &= F(y_k) - F(x^*) - \delta\overline{F}(x_{k-n+1}, x_{k-n})\,(y_k - x^*) \\ &\leqq \{\delta\overline{F}(y_k, x^*) - \delta\overline{F}(x_{k-n+1}, x_{k-n})\}\,(y_k - x^*) \\ &= \{\delta\overline{F}(y_k, x^*) - \delta\overline{F}(x^*, x_{k-n+1}) + \delta\overline{F}(x^*, x_{k-n+1}) - \delta\overline{F}(x_{k-n+1}, x_{k-n})\}\,(y_k - x^*) \end{aligned}$$

and therefore

$$\|y_{k+1} - x^*\| \leqq \delta \cdot \|y_k - x^*\| \cdot (\|y_k - x^*\| + \|x^* - x_{k-n}\|) \ . \tag{5.4}$$

Defining $r_k := \max\{\|x^* - x_k\|, \|y_k - x^*\|\}$, we get by (5.3) and (5.4)

$$r_{k+1} \leqq \max\{\gamma, \delta\}\, r_k(r_k + r_{k-n}) \ . \tag{5.5}$$

5.5) guarantees that the R-order of the sequence (r_k) is not less than τ, where τ is the unique positive root of $t^{n+1} - t^n - 1 = 0$ [3]. It therefore follows that the sequences (x_k), (y_k) also converge with this order.

Remark: It is sufficient for (5.2) that the second partial derivatives of F are bounded [6].

We now consider the computational efficiency of the methods (4.1), (3.6), using Ostrowski's efficiency index, which is defined as $\lambda = \varkappa^{1/\alpha}$, where $\varkappa$ is the order of the iteration and α is the number of function evaluations per iteration step [4]. For (3.6) we need $(n + 2)$ function evaluations per iteration step, for (4.1) 3 (except at the first stage). We therefore get

$$\lambda_{(3.6)} = [(1 + \sqrt{5})/2]^{1/(n+2)} \quad \text{for (3.6) and} \quad \lambda_{(4.1)} = \tau^{\,1/3} \text{for (4.1)}.$$

Table 5.1

k	$\ln \lambda_{(4.1)}$	$\ln \lambda_{(3.6)}$	$\ln \lambda_{(4.1)}/\ln \lambda_{(3.6)}$
3	0.1074 0	0.9624 −1	0.1116 +1
5	0.8364 −1	0.6874 −1	0.1217 +1
10	0.5638 −1	0.4010 −1	0.1406 +1
15	0.4369 −1	0.2831 −1	0.1543 +1
20	0.3612 −1	0.2187 −1	0.1652 +1
50	0.1893 −1	0.9254 −2	0.2046 +1
100	0.1124 −1	0.4718 −2	0.2383 +1

If $A_{(3.6)}$ is the number of function evaluations, which are required to attain a given accuracy by method (3.6) and $A_{(4.1)}$ is defined analogously, then the asymptotic relation $A_{(3.6)}/A_{(4.1)} \approx \ln \lambda_{(4.1)}/\ln \lambda_{(3.6)}$ holds [7].

6. Numerical results

A Gaussian integration rule of order 9 is used for solving the integral equation of Chandrasekhar

$$-H(t) + 1 + \frac{t \cdot H(t)}{4} \cdot \int_0^1 \frac{H(s)}{s + t}\, ds = 0 \,,$$

approximately [5]. This transforms the integral equation into the system of nonlinear equations

$$-H^{(i)} + 1 + \frac{t^{(i)} H^{(i)}}{4} \cdot \sum_{j=1}^{9} \gamma^{(j)} \frac{H^{(j)}}{t^{(j)} + t^{(i)}} = 0 \,, \qquad i = 1(1)\,9 \,, \tag{6.1}$$

where $t^{(i)}$ denote the nodes and $\gamma^{(i)}$ the weights of the integration rule. When solving this system we get an approximate solution $H^4_{(i)}$ of the integral equation at the points $t_{(i)}$.

(6.1) is solved iteratively by (4.1) with the initial points

$$\begin{aligned} &x_{-l+1},\ l = 0(1)\,n\,, \qquad y_1 \\ &x^{(i)}_{-l+1} = \tfrac{1}{2} \cdot \delta_{il} \quad \text{(Kronecker symbol)}\,, \qquad l = 2(1)\,n\,, \\ &x_0^{(i)} = \tfrac{1}{2}\,, \qquad x_1^{(i)} = 1\,, \qquad y_1^{(i)} = 2\,. \end{aligned}$$

It can be shown that the assumptions of the Theorems 4.2 and 5.1 are valid in $[0, y_1]$ with $G = T = I$ and $S = (s_{ij})$,

$$s_{ij} = \begin{cases} 1 - \left(\sum_{k=1}^{9} \gamma^{(k)} - \gamma^{(i)}/4\right)\Big/2\,, & i = j\,, \\ -\gamma^{(j)}/2\,, & i \neq j\,. \end{cases}$$

For $H^{(5)'}$ the approximation of H at the point $t = 0.5$ we get the following enclosing intervals

Table 6.1

Number of Iterations	Enclosing Interval
1	[1.1428708, 1.5201605]
2	[1.1764996, 1.2842208]
3	[1.1849159, 1.2118455]
4	[1.1870473, 1.1933794]
5	[1.1875734, 1.1881804]
6	[1.1877221, 1.1877914]
7	[1.1877336, 1.1877376]
8	[1.1877352, 1.1877352]

The numerical results were obtained on the CD 6500 computer of the Technical University Berlin.

References

1 Kantorovich, L.; The method of successive approximations for functional equations, Acta Math. **71**, 63–97 (1939).
2 Mönch, W.; Monotone Einschließung von Lösungen nichtlinearer Gleichungssysteme durch überlinear konvergente Mehrschrittverfahren, ZAMM **54**, 807–812 (1974).
3 Ortega, J. M.; Rheinboldt, W. C.; Iterative Solutions of Nonlinear Equations in Several Variables, Academic Press 1966.
4 Ostrowski, A.; Solution of Equations and Systems of Equations, Academic Press 1966.
5 Rall, L., Computational Solution of Nonlinear Operator Equations, Wiley 1969.
6 Schmidt, J. W., Eine Übertragung der Regula falsi auf Gleichungen in Banachräumen. I, II, ZAMM **43**, 1–8, 97–100 (1963).
7 Schmidt, J. W., Überlinear konvergente Mehrschrittverfahren vom Regula falsi- und Newton-Typ, ZAMM **53**, 103–114 (1973).
8 Schmidt, J. W.; Leonhardt, H., Eingrenzung von Lösungen mit Hilfe der Regula falsi, Computing **6**, 318–329 (1970).
9 Schneider, N., Monotonie und Einschließung unter Verwendung eines verallgemeinerten Steigungsbegriffes, Dissertation, TU Berlin 1979.
10 Schwetlick, H., Über die Realisierung und Konvergenz von Mehrschrittverfahren zur iterativen Lösung nichtlinearer Gleichungen, ZAMM **54**, 479–493 (1974).
11 Vandergraft, J., Newton's method for convex operators in partially ordered spaces, SIAM J. Numer. Analysis **4**, 406–432 (1967).

Eingereicht am 1. 9. 1980, revidierte Fassung am 14. 3. 1981

Anschrift: Dr. Norbert Schneider, Fachbereich Mathematik, Technische Universität Berlin, Sekretariat H 64, Straße des 17. Juni 135, D-1000 Berlin (West) 12

ZAMM 62, 15 – 25 (1982)

P. Mazilu

Potentialdarstellung einer asymptotischen Lösung der Gleichung $\Delta\Delta u - \varkappa^2 \Delta u = f$

Potentialtheoretische Methoden werden angewendet bei der Untersuchung der singulären Störungen der harmonischen und biharmonischen Differentialgleichung. Mit Hilfe dieser Methoden kann die Johnsche Grenzschichtabschätzung für homogene Gleichungen erhalten werden. Die harmonische Funktion, die man bei dieser Abschätzung findet, wird bestimmt durch eine Funktionalgleichung im Raum der harmonischen Funktionen. Außerdem wird auch die Ausdehnung dieses Ergebnisses auf inhomogene Gleichungen möglich.

Potential theoretic methods are applied to solve singular perturbation problems for harmonic and biharmonic equations. Using these methods John's boundary layer evaluation formula for the homogeneous equations can be derived. The harmonic function encountered in this formula is determined by a functional equation defined in the space of harmonic functions. Moreover it is also possible to derive similar results for non-homogeneous equations.

Применяются методы теории потенциала крешению случая сингулярных возмущений в гармонических и бигармонических уравнениях. При использовании этой методы получается и оценка формулы для пограничного слоя на примере однородного уравнения. Первая формула для этой задачи была получена Ионом. В настоящей работе кроме того предлагается обобщение этой формулы для неоднородных уравнений.

1. Einleitung

Die Gleichung $\varepsilon^2 \Delta\Delta u - \Delta u = f$ (oder äquivalent dazu $\Delta\Delta u - \varkappa^2 \Delta u = \varkappa^2 f$) mit der Standard-Randwertaufgabe $u|_{\partial D} = g$, $(\partial u/\partial n)|_{\partial D} = h$, wobei ε oder $\varkappa$ kleine bzw. große reellwertige Parameter sind, wird untersucht. Sie tritt z. B. in der Theorie elastischer Platten auf.

Wie F. John [1] gezeigt hat, existiert unter der speziellen Bedingung $f = 0$ eine harmonische Funktion w, die sich außerhalb einer Grenzschicht gut der Lösung u nähert. Diese Näherung lautet

$$u(P) = w(P) + O(\mathrm{e}^{-(d/2\varepsilon)})$$

wobei d den Abstand von P zum Rand ∂D bezeichnet.

Der komplementäre Fall, $g = h = 0$ und $f \neq 0$, wurde von Friedmann [2], Greenlee [3], Lions [4] und Schuss [5] untersucht. Die wesentlichen Ergebnisse dieser Autoren sind der Nachweis der L_2-Konvergenz der Lösung und ihrer Ableitungen erster Ordnung gegen die Lösung der Poissonschen Gleichung $\Delta w = f$ mit Dirichletschen Randbedingungen $w|_{\partial D} = 0$ und deren entsprechenden Ableitungen.

In der vorliegenden Arbeit soll die oben erwähnte Randwertaufgabe mit Hilfe potentialtheoretischer Methoden untersucht werden. Hierzu ersetzen wir die Originalgleichung durch das Gleichungssystem

$$\Delta u - \varkappa^2 u = F + w\,, \qquad \Delta w = 0\,,$$

wobei F die Lösung der Poissonschen Gleichung $\Delta F = f$ mit homogenen Dirichletschen Randbedingungen ist. Die harmonische Funktion w wird bestimmt durch eine Funktionalgleichung vom Typus

$$(w, \varphi) = F(\varphi)\,,$$

wobei (,) ein spezielles Skalarprodukt und $F(\cdot)$ ein bestimmtes Funktional im Raum der harmonischen Funktionen ist.

Die Lösung u können wir durch eine Potentialdarstellung ausdrücken, wobei F, w und die Randwerte g und h Dichten sind. Auf diesem Weg ist für homogene Gleichungen ($f = 0$ und $g \neq 0$, $h \neq 0$) eine Bestimmung der Johnschen Abschätzung möglich. Man erhält nämlich

$$u(P) = w(P) + O\left(\frac{1}{\varepsilon^2} \sqrt{\frac{\varepsilon}{d}}\; \mathrm{e}^{-(d/\varepsilon)}\right).$$

Für die komplementäre Randwertaufgabe ($f \neq 0$ und $g = h = 0$) erhält man die Abschätzung

$$u(P) = -\frac{1}{2\pi\varepsilon^2} \int\limits_D F(Q)\, K_0\left(\frac{1}{\varepsilon}\, r_{PQ}\right) \mathrm{d}x\, \mathrm{d}y - \varepsilon^2 w(P) + O\left(\frac{1}{\varepsilon^2} \sqrt{\frac{\varepsilon}{d}}\; \mathrm{e}^{-(d/\varepsilon)}\right),$$

wobei K_0 die Hankelsche Funktion bezeichnet. Diese Abschätzung zeigt, daß die von den homogenen Gleichungen bestimmte Grenzschicht auch für den inhomogenen Fall existiert.

2. Fundamentallösung für die Gleichung $\Delta u - \varkappa^2 u = 0$

Betrachten wir die Gleichung

$$\Delta u - \varkappa^2 u = 0\,, \tag{2.1}$$

wobei Δ den Laplaceschen Differentialoperator bezeichnet und $\varkappa$ eine reelle Zahl. Die Fundamentallösung der Gleichung (2.1) kann durch die Hankelsche Funktion

$$K_0(r) = \int\limits_0^\infty \mathrm{e}^{-r \cosh \eta}\, \mathrm{d}\eta \tag{2.2}$$

definiert werden.

Zunächst werden einige Eigenschaften von $K_0(r)$ betrachtet:

1. *Im Punkt $r = 0$ weist $K_0(r)$ eine logarithmische Singularität*

$$K_0(r) = \ln\frac{1}{r} + \dots \tag{2.3}$$

auf, wobei die Punkte diejenigen Glieder bezeichnen, die für $r \to 0$ beschränkt bleiben.

2. *Für große Werte von r lautet die asymptotische Darstellung*

$$K_0(r) = \sqrt{\frac{\pi}{2r}}\,e^{-r}\,(1 + \eta(r)) \tag{2.4}$$

mit

$$\lim_{r\to\infty}\eta(r) = 0 .$$

3. *Für große Werte von r erhält man für die Ableitung $K_0'(r)$ die asymptotische Darstellung*

$$K_0'(r) = -\sqrt{\frac{\pi}{2r}}\,e^{-r}\,(1 + \eta(r)) \tag{2.5}$$

mit

$$\lim_{r\to\infty}\eta(r) = 0 .$$

4. *Die Ableitung $K_0'(r)$ weist im Punkt $r = 0$ eine polare Singularität*

$$K_0'(r) = -\frac{1}{r}\,\Phi(r) = -\frac{1}{r} + \dots \tag{2.6}$$

auf, wobei Punkte diejenigen Glieder bezeichnen, die für $r = 0$ beschränkt bleiben.

Die Beweise der ersten zwei Eigenschaften wurden in [6] entwickelt. Gleichartige Methoden kann man anwenden, um die Formeln (2.5) und (2.6) zu beweisen.

Ein anderer Weg, die oben erwähnten Eigenschaften zu beweisen, grundet sich auf die Formeln für die asymptotische Darstellung von Parameter-Integralen, die in einem allgemeinen mathematischen Rahmen in dem Buch von BERG [15] betrachtet werden.

Es sei P ein fester Punkt aus der Ebene (x, y). Dann ist die Funktion

$$g(P, Q) = K_0(\varkappa r_{PQ})$$

mit

$$r_{PQ} \doteq \sqrt{(x_Q - x_P)^2 + (y_Q - y_P)^2}$$

fur festes P und für $Q \neq P$ eine Lösung der Gleichung (2.1). Diese Lösung hat dieselben Eigenschaften wie die Fundamentallösung $\ln(1/r_{PQ})$ der LAPLACEschen Gleichung. Im folgenden werden wir diese Analogie benutzen.

3. Potentialdarstellung

Es sei D ein beschränktes Gebiet aus der Ebene (x, y), das von einer hinreichend glatten Kurve ∂D begrenzt ist. Betrachten wir nun den Differentialoperator

$$L(u) = \Delta u - \varkappa^2 u , \tag{3.1}$$

wobei u eine beliebige Funktion aus $\mathcal{C}^2(D)$ ist. Dann gilt die Identität

$$\int_D (L(u)\,v - uL(v))\,dx\,dy = \int_{\partial D}\left(\frac{\partial u}{\partial n}v - u\frac{\partial v}{\partial n}\right)ds \tag{3.2}$$

fur beliebige u, v aus $\mathcal{C}^2(D) \cap \mathcal{C}^1(D + \partial D)$ ($\partial/\partial n$ bezeichnet die äußere Normalableitung).

Wenn wir $K_0(\varkappa r_{PQ})$ als Funktion v annehmen und als Gebiet $D - D_\varepsilon(P)$ wählen, wobei $D_\varepsilon(P)$ den Kreis mit dem Radius ε und dem Mittelpunkt P darstellt, dann erhalten wir anstatt (3.2)

$$\int_{D-D_\varepsilon} L(u)\,K_0(r_{PQ})\,dx\,dy = \int_{\partial D}\left(\frac{\partial u}{\partial n}K_0(\varkappa r_{PQ}) - u\frac{\partial K_0}{\partial n}(\varkappa r_{PQ})\right)ds + \int_{\partial D_\varepsilon(P)}\left(\frac{\partial u}{\partial n}K_0(\varkappa r_{PQ}) - u\frac{\partial K_0}{\partial n}(\varkappa r_{PQ})\right)ds . \tag{3.3}$$

Aufgrund der Eigenschaften 1 und 3 von $K_0(r)$ ergibt sich

$$\lim_{\varepsilon\to 0}\int_{\partial D_\varepsilon(P)}\frac{\partial u}{\partial n}K_0(\varkappa r_{PQ})\,ds = 0$$

und

$$\lim_{\varepsilon\to 0}\int_{\partial D_\varepsilon(P)} u\frac{\partial K_0}{\partial n}(\varkappa r_{PQ})\,ds = 2\pi u(P) .$$

Läßt man ε in (3.3) gegen Null streben, so erhält man die Integraldarstellung

$$u(P) = -\frac{1}{2\pi} \int\limits_D L(u)\, K_0(\varkappa r_{PQ})\, dx\, dy + \frac{1}{2\pi} \int\limits_{\partial D} \left(\frac{\partial u}{\partial n} K_0(\varkappa r_{PQ}) - u \frac{\partial K_0}{\partial n}(\varkappa r_{PQ})\right) ds . \tag{3.4}$$

Falls u eine Lösung der Differentialgleichung $L(u) = f$ ist, mit f als gegebener stetiger Funktion, dann lautet

$$u(P) = -\frac{1}{2\pi} \int\limits_D f(Q)\, K_0(\varkappa r_{PQ})\, dx\, dy + \int\limits_{\partial D} \left(\frac{\partial u}{\partial n} K_0(\varkappa r_{PQ}) - u \frac{\partial K_0}{\partial n}(\varkappa r_{PQ})\right) ds . \tag{3.5}$$

Dies liefert als Lösung der homogenen Differentialgleichung $L(u^*) = 0$

$$u^*(P) = \frac{1}{2\pi} \int\limits_{\partial D} \left(\frac{\partial u}{\partial n} K_0(\varkappa r_{PQ}) - u \frac{\partial K_0}{\partial n}(\varkappa r_{PQ})\right) ds . \tag{3.6}$$

4. Asymptotisches Verhalten

Bezeichnen wir mit $d = \min\limits_{Q \in \partial D} r_{PQ}$ die Entfernung von P zu ∂D, dann gilt

$$u^*(P) = O\left(\frac{1+\varkappa}{\sqrt{\varkappa d}} e^{-\varkappa d}\right), \tag{4.1}$$

($O(\varepsilon)$ bezeichnet eine Funktion, die für hinreichend kleine ε der Bedingung $|O(\varepsilon)| \leqq k\varepsilon$ genugt, wobei k eine feste Konstante ist).

Zum Beweis betrachten wir die Integrale

$$I_1 = \int\limits_{\partial D} \frac{\partial u}{\partial n} K_0(\varkappa r_{PQ})\, ds , \qquad I_2 = \int\limits_{\partial D} u \frac{\partial K_0}{\partial n}(\varkappa r_{PQ})\, ds$$

und die aus ihnen resultierenden Ungleichungen

$$|I_1| \leqq \max_{Q \in \partial D} \left|\frac{\partial u^*}{\partial n}(Q)\right| l \max_{Q \in \partial D} |K_0(\varkappa r_{PQ})| , \qquad |I_2| \leqq \max_{Q \in \partial D} |u^*(Q)|\, l \max_{Q \in \partial D} |K_0(\varkappa r_{PQ})| ,$$

wobei l die Länge von ∂D bezeichnet. Aus der Eigenschaft (2.4) der Funktion $K_0(r)$ folgt

$$\max_{Q \in \partial D} |K_0(\varkappa r_{PQ})| = \max_{Q \in \partial D} O\left(\frac{e^{-\varkappa r_{PQ}}}{\sqrt{\varkappa r_{PQ}}}\right) = O\left(\frac{e^{-\varkappa d}}{\sqrt{\varkappa d}}\right)$$

und

$$\max_{Q \in \partial D} \left|\frac{\partial K_0}{\partial n}(\varkappa r_{PQ})\right| = \varkappa \max_{Q \in \partial D} \left|O\left(\frac{e^{-\varkappa r_{PQ}}}{\sqrt{\varkappa r_{PQ}}}\right)\right| = O\left(\varkappa \frac{e^{-\varkappa d}}{\sqrt{\varkappa d}}\right).$$

Damit ist

$$|I_1| \leqq \left(\max_{Q \in \partial D} \left|\frac{\partial u^*}{\partial n}\right|\right) l\, O\left(\frac{e^{-\varkappa d}}{\sqrt{\varkappa d}}\right) \tag{4.2}$$

und

$$|I_2| \leqq \left(\max_{Q \in \partial D} |u^*|\right) l\, O\left(\varkappa \frac{e^{-\varkappa d}}{\sqrt{\varkappa d}}\right), \tag{4.3}$$

womit (4.1) bewiesen ist.

Anmerkung: Da in den Beziehungen (4.2) und (4.3) die Glieder $\max |u^*|$ und $\max |\partial u^*/\partial n|$ auftreten, kann (4.1) nicht als eine asymptotische Abschätzung für eine Lösung der Dirichletschen oder Neumannschen Randbedingung benutzt werden.

Jetzt betrachten wir die inhomogene Gleichung

$$L(u) = f , \tag{4.4}$$

wobei $f(x, y)$ eine in dem Gebiet D definierte gegebene Funktion ist.

Zu beweisen sind die folgenden Sätze:

a) *Falls $f \in \mathscr{C}^2(D) \cap \mathscr{C}^0(D + \partial D)$, dann gilt innerhalb von D*

$$\varkappa^2 u(P) = -f(P) + O\left(\frac{1}{\varkappa^2}\right). \tag{4.5}$$

b) *Wenn $f \in \mathscr{C}^3(D) \cap \mathscr{C}^0(D + \partial D)$, dann gilt innerhalb von D*

$$\varkappa^2 \frac{\partial u}{\partial x} = -\frac{\partial f}{\partial x} + O\left(\frac{1}{\varkappa^2}\right), \qquad \varkappa^2 \frac{\partial u}{\partial y} = -\frac{\partial f}{\partial y} + O\left(\frac{1}{\varkappa^2}\right). \tag{4.6), (4.7}$$

Zum Beweis stellen wir die Lösung von (4.4) in der Form

$$u = u_0 + u^*$$

dar, mit

$$u_0 = -\frac{1}{2\pi} \int_D f(Q)\, K_0(\varkappa r_{PQ})\, \mathrm{d}x\, \mathrm{d}y$$

und u^* als Lösung der homogenen Gleichung $L(u^*) = 0$.

Aufgrund von (4.1) erhält man

$$u^* = O\left(\frac{1+\varkappa}{\sqrt{\varkappa d}} \mathrm{e}^{-\varkappa d}\right)$$

und somit

$$\varkappa^2 u^* = O\left(\frac{\varkappa^2(1+\varkappa)}{\sqrt{\varkappa d}} \mathrm{e}^{-\varkappa d}\right) = O\left(\frac{1}{\varkappa^2}\right). \tag{4.8}$$

Zum Beweis der analogen Beziehung für u_0 schreiben wir die Integraldarstellung von u_0 in der Form

$$u_0(P) = -\frac{1}{2\pi} \int_{D - D_d(P)} f(Q)\, K_0(\varkappa r_{PQ})\, \mathrm{d}x\, \mathrm{d}y - \frac{1}{2\pi} \int_{D_d(P)} f(Q)\, K_0(\varkappa r_{PQ})\, \mathrm{d}x\, \mathrm{d}y\,,$$

wobei $D_d(P)$ der Kreis mit dem Radius d und dem Mittelpunkt P ist. Aufgrund der Eigenschaften von $K_0(r)$ genügt das erste der Integrale der Ungleichung

$$\left| \int_{D - D_d(P)} f(Q)\, K_0(\varkappa r_{PQ})\, \mathrm{d}x\, \mathrm{d}y \right| \leqq \max_{Q \in D} |f(Q)|\, |K_0(\varkappa d)|\, m(D) \leqq$$

$$\leqq \sqrt{\frac{\pi}{2}} \max |f(Q)|\, m(D) \frac{\mathrm{e}^{-\varkappa d}}{\sqrt{\varkappa d}} (1 + \eta(\varkappa d)) = O\left(\frac{\mathrm{e}^{-\varkappa d}}{\sqrt{\varkappa d}}\right),$$

$m(D)$ bezeichnet dabei den Inhalt des Gebietes D. Wendet man eine Taylor-Entwicklung für das zweite Integral an, so folgt

$$\int_{D_d(P)} f(Q)\, K_0(\varkappa r_{PQ})\, \mathrm{d}x\, \mathrm{d}y = f(P) \int_{D_d(P)} K_0(\varkappa r_{PQ})\, \mathrm{d}x\, \mathrm{d}y + \frac{\partial f}{\partial x}(P) \int_{D_d(P)} (x_Q - x_P)\, K_0(\varkappa r_{PQ})\, \mathrm{d}x\, \mathrm{d}y +$$

$$+ \frac{\partial f}{\partial y}(P) \int_{D_d(P)} (y_Q - y_P)\, K_0(\varkappa r_{PQ})\, \mathrm{d}x\, \mathrm{d}y + \int_{D_d(P)} O(r_{PQ}^2)\, K_0(\varkappa r_{PQ})\, \mathrm{d}x\, \mathrm{d}y\,.$$

Nun betrachten wir der Reihe nach die Integrale der letzten Beziehung. Für das erste ergibt sich

$$\int_{D_d(P)} K_0(\varkappa r_{PQ})\, \mathrm{d}x\, \mathrm{d}y = 2\pi \int_0^d \varrho K_0(\varkappa\varrho)\, \mathrm{d}\varrho = \frac{2\pi}{\varkappa^2} \int_0^{\varkappa d} r K_0(r)\, \mathrm{d}r = \frac{2\pi}{\varkappa^2} \int_0^{\varkappa d} \frac{\mathrm{d}}{\mathrm{d}r}\left(r \frac{\mathrm{d}K_0}{\mathrm{d}r}(r)\right) \mathrm{d}r =$$

$$= \frac{2\pi}{\varkappa^2}\left(\varkappa d \frac{\mathrm{d}K_0}{\mathrm{d}r}(\varkappa d) - r \frac{\mathrm{d}K_0}{\mathrm{d}r}(r)\Big|_{r=0}\right) = \frac{2\pi}{\varkappa^2}\left(1 + \varkappa d\, O\left(\frac{1}{\sqrt{\varkappa d}} \mathrm{e}^{-\varkappa d}\right)\right) =$$

$$= \frac{2\pi}{\varkappa^2}\left(1 + O\left(\sqrt{\varkappa d}\, \mathrm{e}^{-\varkappa d}\right)\right).$$

Für die folgenden zwei erhält man

$$\int_{D_d(P)} (x_Q - x_P)\, K_0(\varkappa r_{PQ})\, \mathrm{d}x\, \mathrm{d}y = \int_0^{2\pi} \cos\theta\, \mathrm{d}\theta \int_0^d \varrho^2 K_0(\varkappa\varrho)\, \mathrm{d}\varrho = 0$$

und

$$\int_{D_d(P)} (y_Q - y_P)\, K_0(\varkappa r_{PQ})\, \mathrm{d}x\, \mathrm{d}y = \int_0^{2\pi} \sin\theta\, \mathrm{d}\theta \int_0^d \varrho^2 K_0(\varkappa\varrho)\, \mathrm{d}\varrho = 0\,.$$

Das letzte Integral wird zu

$$\int_{D_d(P)} O(r_{PQ}^2)\, K_0(\varkappa r_{PQ})\, \mathrm{d}x\, \mathrm{d}y = 2\pi \int_0^d O(\varrho^3)\, K_0(\varkappa\varrho)\, \mathrm{d}\varrho = \frac{2\pi}{\varkappa^4} \int_0^{\varkappa d} O(r^3)\, K_0(r)\, \mathrm{d}r = O\left(\frac{1}{\varkappa^4}\right),$$

und damit gilt

$$\int_{D_d(P)} f(Q)\, K_0(\varkappa r_{PQ})\, \mathrm{d}x\, \mathrm{d}y = \frac{2\pi}{\varkappa^2} f(P) + O\left(\frac{1}{\varkappa^4}\right) + O\left(\sqrt{\varkappa d}\, \mathrm{e}^{-\varkappa d}\right)$$

und somit

$$\varkappa^2 u_0(\varrho) = -f(P) + O\left(\frac{1}{\varkappa^2}\right) + O\left(\varkappa^2 \sqrt{\varkappa d}\, e^{-\varkappa d}\right) + O\left(\frac{\varkappa^2}{\sqrt{\varkappa d}} e^{-\varkappa d}\right) = -f(P) + O\left(\frac{1}{\varkappa^2}\right), \tag{4.9}$$

woraus zusammen mit (4.8) die Beziehung (4.5) folgt. Der Beweis von (4.6) oder (4.7) läuft entsprechend.

Anmerkung: Als alternativen Beweis von (4.6) verwendet man, allerdings unter der Voraussetzung

$$f \in \mathscr{C}^3(D) \cap \mathscr{C}^1(D + \partial D),$$

die Identitäten

$$\varkappa^2 \frac{\partial u}{\partial x}(P) = -\frac{1}{2\pi} \int_D f(Q) \frac{\partial}{\partial x_P} \left(K_0(\varkappa r_{PQ})\right) dx\, dy =$$

$$= -\frac{1}{2\pi} \int_D \frac{\partial}{\partial x_Q} \left(f(Q)\right) K_0(\varkappa r_{PQ})\, dx\, dy + \frac{1}{2\pi} \int_{\partial D} f(Q)\, K_0(\varkappa r_{PQ})\, n_x\, ds$$

und

$$\varkappa^2 \frac{\partial u}{\partial y}(P) = -\frac{1}{2\pi} \int_D f(Q) \frac{\partial}{\partial y_P} \left(K_0(\varkappa r_{PQ})\right) dx\, dy =$$

$$= -\frac{1}{2\pi} \int_D \frac{\partial}{\partial y_Q} \left(f(Q)\right) K_0(\varkappa r_{PQ})\, dx\, dy + \frac{1}{2\pi} \int_{\partial D} f(Q)\, K_0(\varkappa r_{PQ})\, n_y\, ds$$

mit n_x und n_y als Projektion auf die Koordinatenachsen. So erhält man wieder die vorher erwähnten Integrale.

5. Integralgleichungen

Gesucht ist die Lösung der Gleichung (2.1)

$$u(P) = -\int_{\partial D} \nu(Q) \frac{\partial K_0}{\partial n_Q}(\varkappa r_{PQ})\, ds_Q \tag{5.1}$$

oder

$$u(P) = \int_{\partial D} \mu(Q)\, K_0(\varkappa r_{PQ})\, ds_Q. \tag{5.2}$$

Berücksichtigen wir die Eigenschaften 1 und 2 der Funktion $K_0(r)$, so ist es uns, wie gewöhnlich in der Potentialtheorie, erlaubt, auf der Berandung

$$\pm \pi\nu(P) - \int_{\partial D} \nu(Q) \frac{\partial K_0}{\partial n_Q}(\varkappa r_{PQ})\, ds_Q = (u)^{\pm}, \tag{5.3}$$

$$\mp \pi\mu(P) + \int_{\partial D} \mu(Q) \frac{\partial K_0}{\partial n_P}(\varkappa r_{PQ})\, ds_Q = \left(\frac{\partial u}{\partial n}\right)^{\pm} \tag{5.4}$$

anzunehmen, wobei $(\ )^+$ die inneren Randwerte und $(\ )^-$ die äußeren Randwerte (in bezug auf D) bezeichnen. Wir betrachten die Kerne

$$K(P, Q) = -\frac{\partial K_0}{\partial n_Q}(\varkappa r_{PQ}) \quad \text{und} \quad K^*(P, Q) = \frac{\partial K_0}{\partial n_P}(\varkappa r_{PQ}).$$

Wenn man mit φ und ψ die Winkel zwischen den Vektoren $\overline{PQ}$ und die inneren (in bezug auf D) Normalen an ∂D in Q bzw. in P bezeichnet, dann gilt

$$K(P, Q) = \varkappa K_0'(\varkappa r_{PQ}) \cos\varphi$$

und

$$K^*(P, Q) = \varkappa K_0'(\varkappa r_{PQ}) \cos\psi.$$

Wir bemerken, daß $K(P, Q)$ und $K^*(P, Q)$ adjungiert sind, d. h.

$$K^*(P, Q) = K(Q, P).$$

Demzufolge ist es möglich, die FREDHOLMsche Alternative anzuwenden.

Aufgrund der Eigenschaft (2.6) für $K_0'(r)$ wird

$$K(P, Q) = \frac{\cos\varphi}{r_{PQ}} \Phi(\varkappa r_{PQ})$$

und

$$K^*(P, Q) = \frac{\cos\psi}{r_{PQ}} \Phi(\varkappa r_{PQ}),$$

wobei

$$\Phi(r) = \int_r^\infty \frac{\lambda\, e^{-\lambda}}{\sqrt{\lambda^2 - r^2}}\, d\lambda .$$

Es ist möglich zu beweisen daß diese Funktion die folgenden Eigenschaften hat:

a) $\lim_{r\to\infty} \Phi(r) = 1$,

b) *fur große Werte von r gilt*

$$\Phi(r) = O(\sqrt{r}\, e^{-r}),$$

c) *Φ ist eine beschränkte Funktion*

$$\max_{r\in[0,\infty)} |\Phi(r)| = M < +\infty .$$

Jetzt betrachten wir die Integraloperatoren

$$I(\nu) = \frac{1}{\pi} \int_{\partial D} K(P, Q)\, \nu(Q)\, ds \tag{5.5}$$

und

$$I^*(\mu) = \frac{1}{\pi} \int_{\partial D} K^*(P, Q)\, \mu(Q)\, ds . \tag{5.6}$$

Wir werden den folgenden Satz beweisen: *Falls ∂D stetige Krümmung hat und $1/\varkappa$ hinreichend klein ist, so definieren $I(\nu)$ und $I^*(\mu)$ Kontraktionen, d. h. es gibt $0 < k < 1$, so daß*

$$\max_{Q\in\partial D} |I(\nu)| \leqq k \max_{Q\in\partial D} |\nu| \tag{5.7}$$

und

$$\max_{Q\in\partial D} |I^*(\mu)| \leqq k \max |\mu| . \tag{5.8}$$

Wir beweisen diesen Satz für $I(\nu)$. Zu diesem Zweck stellen wir $I(\nu)$ in der Form

$$I(\nu) = \int_{C_\varepsilon} \nu(Q) \frac{\cos\varphi}{r_{PQ}} \Phi(\varkappa r_{PQ})\, ds + \int_{\partial D - C_\varepsilon} \nu(Q) \frac{\cos\varphi}{r_{PQ}} \Phi(\varkappa r_{PQ})\, ds$$

dar, wobei C_ε der Teil von ∂D ist, der in dem Kreis mit dem Radius ε und dem Mittelpunkt P liegt. Wir erhalten

$$\max_{Q\in\partial D} |I(\nu)| \leqq O(\varepsilon) \max_{Q\in\partial D} \left|\frac{\cos\varphi}{r_{PQ}}\right| \max_{Q\in\partial D} |\nu| + O(\sqrt{\varkappa\varepsilon}\, e^{-\varkappa\varepsilon}) \max_{Q\in\partial D} \left|\frac{\cos\varphi}{r_{PQ}}\right| \max_{Q\in\partial D} |\nu| . \tag{5.9}$$

Bestimmen wir jetzt ε und $\varkappa$, so daß der Ausdruck

$$\max \left|\frac{\cos\varphi}{r_{PQ}}\right| \left(O(\varepsilon) + O(\sqrt{\varkappa\varepsilon}\, e^{-\varkappa\varepsilon})\right) = k < 1$$

wird, dann folgt aus (5.9) die Beziehung (5.7).

Anmerkung: Der letzte Satz gilt auch, wenn die Krummung von ∂D nur beschränkt und eine stückweise stetige Funktion ist.

Weil $I(\nu)$ und $I^*(\mu)$ Kontraktionen sind, ist die Lösung der Integralgleichungen aus (5.3) und (5.4) durch sukzessive Approximation bestimmbar:

$$\nu_0 = \pm \frac{1}{\pi} g, \qquad \nu_n = \pm \frac{1}{\pi} g \pm I(\nu_{n-1})$$

und

$$\mu_0 = \mp \frac{1}{\pi} h, \qquad \mu_n = \mp \frac{1}{\pi} h \pm I^*(\nu_{n-1}),$$

wobei $g(s)$ und $h(s)$ die Randwerte für die DIRICHLETsche bzw. NEUMANNsche Randwertaufgabe sind. Offenbar streben die Reihen $\{\nu_n\}_{n\in N}$ und $\{\mu_n\}_{n\in N}$ gegen die Lösungen ν bzw. μ der Lösungen (5.3) und (5.4). Die Fehler werden dann

$$\max_{Q\in\partial D} |\nu_n - \nu| < \frac{k^n}{1-k} \max_{Q\in\partial D} |g|, \qquad \max_{Q\in\partial D} |\mu_n - \mu| < \frac{k^n}{1-k} \max_{Q\in\partial D} |h| . \tag{5.10), (5.11}$$

Ebenfalls gelten die folgenden Abschätzungen

$$\max_{Q\in\partial D} |\nu| < \frac{1}{\pi(1-k)} \max |g| \tag{5.12}$$

und

$$\max_{Q \in \partial D} |\mu| < \frac{1}{\pi(1-k)} \max_{Q \in \partial D} |h| \, . \tag{5.13}$$

Jetzt können wir noch eine a priori asymptotische Abschätzung für die Dirichletsche oder Neumannsche Randwertaufgabe bestimmen. Aus (5.1) und (5.2) folgt

$$|u(P)| \leqq \max_{Q \in \partial D} |\nu| \, O\left(\frac{e^{-\varkappa d}}{\sqrt{\varkappa d}}\right)$$

und

$$|u(P)| \leqq \max_{Q \in \partial D} |\mu| \, O\left(\frac{e^{-\varkappa d}}{\sqrt{\varkappa d}}\right).$$

Hieraus folgt

$$|u(P)| \leqq \frac{1}{\pi(1-k)} \max_{Q \in \partial D} |g| \, O\left(\frac{e^{-\varkappa d}}{\sqrt{\varkappa d}}\right)$$

und

$$|u(P)| \leqq \frac{1}{\pi(1-k)} \max_{Q \in \partial D} |h| \, O\left(\frac{e^{-\varkappa d}}{\sqrt{\varkappa d}}\right).$$

Anmerkung: Die Methode der Potentialdarstellung und Integralgleichung kann für die numerische Bestimmung der asymptotischen Lösungen der Gleichung (2.1) benutzt werden. Wenn man die Glieder der Ordnung $O(1/\varkappa)$ vernachlässigt, dann sind die Lösungen der Dirichletschen und Neumannschen Randwertaufgabe:

$$u(P) = \pm \frac{1}{\pi} \int_{\partial D} g(P) \frac{\partial K_0}{\partial n_Q} (\varkappa r_{PQ}) \, ds$$

bzw.

$$u(P) = \mp \int_{\partial D} h(Q) \, K_0(\varkappa r_{PQ}) \, ds \, .$$

Diese Methode gibt eine gute Approximation auch für den Fall, daß die gewöhnlichen Methoden nicht mehr funktionieren (z. B. wenn g oder h schnell oszillierende Funktionen sind).

6. Fundamental-Randwertaufgabe für die homogene Gleichung $\Delta\Delta u - \varkappa^2 \Delta u = 0$

Wir betrachten die folgenden Randwertaufgaben: Man bestimme $u \in \mathscr{C}^4(D) \cap \mathscr{C}^1(D + \partial D)$ als Lösung der Gleichung

$$\Delta\Delta u - \varkappa^2 \Delta u = 0 \, , \tag{6.1}$$

so daß auf dem Rand ∂D gilt

$$u|_{\partial D} = g(s) \, , \qquad \left.\frac{\partial u}{\partial n}\right|_{\partial D} = h(s) \, , \tag{6.2), (6.3}$$

wobei $g(s)$ und $h(s)$ gegebene Funktionen sind.

Da es immer möglich ist, eine im Gebiet D harmonische Funktion u^* so zu bestimmen, daß

$$u^*|_{\partial D} = g(s)$$

gilt, kann man nach einer Translation $u \to u^* + u$ in der Randbedingung (6.2) immer annehmen (ohne Beschränkung der Allgemeinheit), daß $g = 0$ ist.

Es sei w die in D harmonische Funktion

$$w = \Delta u - \varkappa^2 u \, . \tag{6.4}$$

Aufgrund des Ergebnisses von § 3 erhalten wir

$$u(P) = -\frac{1}{2\pi} \int_{D} w(Q) \, K_0(\varkappa k_{PQ}) \, dx \, dy + \frac{1}{2\pi} \int_{\partial D} h(Q) \, K_0(\varkappa r_{PQ}) \, ds \, . \tag{6.5}$$

Das heißt, daß dann, wenn die harmonische Funktion w bestimmt ist, die Lösung der Fundamental-Randwertaufgabe auch bestimmt ist. Wir werden eine Funktionalgleichung betrachten, die eine direkte Bestimmung von w gestattet.

Es sei $\mathscr{H}(D)$ der Raum von den in D harmonischen Funktionen, die stetig in $D + \partial D$ sind

$$\mathscr{H}(D) = \{w \in \mathscr{C}^2(D) \cap \mathscr{C}^0(D + \partial D) \mid \Delta w = 0\} \, .$$

Zu jeder Funktion $w \in \mathscr{H}(D)$ betrachten wir die Lösung $u \in \mathscr{C}^2(D) \cap \mathscr{C}^0(D + \partial D)$ der Randwertaufgabe

$$\Delta u - \varkappa^2 u = w \, , \quad u|_{\partial D} = 0 \, . \tag{6.6}$$

Aus der Identität

$$\int_D (|\Delta u|^2 + 2\varkappa^2 |\mathrm{grad}\, u|^2 + \varkappa^4 |u|^2)\, dx\, dy = \int_D |w|^2\, dx\, dy$$

ergibt sich

$$||u||_{L_2}^2 \equiv \int_D |u|^2\, dx\, dy \leqq \frac{1}{\varkappa^4} \int_D |w|^2\, dx\, dy \equiv \frac{1}{\varkappa^4} ||w||_{L_2}^2 ,$$

$$||\mathrm{grad}\, u||_{L_2}^2 = \int_D |\mathrm{grad}\, u|^2\, dx\, dy \leqq \frac{1}{2\varkappa^2} \int_D |w|^2\, dx\, dy = \frac{1}{2\varkappa^2} ||w||_{L_2}^2 ,$$

$$||\Delta u||_{L_2}^2 = \int_D |\Delta u|^2\, dx\, dy \leqq \int_D |w|^2\, dx\, dy = ||w||_{L_2}^2 ,$$

und damit

$$\int_D (|\Delta u|^2 + \varkappa^2 |\mathrm{grad}\, u|^2)\, dx\, dy \leqq \tfrac{3}{2} ||w||_{L_2}^2 . \tag{6.7}$$

Jetzt bestimmen wir eine umgekehrte Ungleichung, und zwar

$$||w||_{L_2}^2 \leqq O(\varkappa^2) \int_D (|\Delta u|^2 + \varkappa^2 |\mathrm{grad}\, u|^2)\, dx\, dy . \tag{6.8}$$

Zunächst berücksichtigen wir, daß

$$||u||_{L_2} \leqq k_1 ||\mathrm{grad}\, u||_{L_2} ,$$

wobei k_1 eine Konstante ist, die nur vom Gebiet D abhängt. Somit ergibt sich aus (6.6)

$$||w||_{L_2} \leqq ||\Delta u||_{L_2} + k_1 \varkappa^2 ||\mathrm{grad}\, u||_{L_2} \leqq (1 + k_1 \varkappa) \left(\int_D (|\Delta u|^2 + \varkappa^2 |\mathrm{grad}\, u|^2) \right)^{1/2} dx\, dy ,$$

was zu beweisen war.

Betrachten wir jetzt in $\mathcal{H}(D)$ das Skalarprodukt und die Norm

$$\langle w, \varphi \rangle = \int_D (\Delta u\, \Delta v + \varkappa^2\, \mathrm{grad}\, u\, \mathrm{grad}\, v)\, dx\, dy ,$$

bzw.

$$||w||_{\mathcal{H}} = \left(\int_D (|\Delta u|^2 + \varkappa^2 |\mathrm{grad}\, u|^2)\, dx\, dy \right)^{1/2} ,$$

wobei u und v durch die Randwertaufgabe (6.6) zu bestimmen sind. Aufgrund der Ungleichungen (6.7) und (6.8) folgt

$$\tfrac{2}{3} ||w||_{\mathcal{H}} \leqq ||w||_{L_2} \leqq O(\varkappa)\, ||w||_{\mathcal{H}} \tag{6.9}$$

und direkt aus der Normdefinition

$$||\mathrm{grad}\, u||_{L_2} \leqq \frac{1}{\varkappa} ||w||_{\mathcal{H}} . \tag{6.10}$$

Mit $\overline{\mathcal{H}}(D)$ bezeichnet man die Vervollständigung von $\mathcal{H}(D)$ in bezug auf die Norm $||\ \ ||_{\mathcal{H}}$.

Wir beweisen den folgenden Satz

Theorem: *Falls die Lösung u der Randwertaufgabe* (6.1-3) *(mit $g = 0$) die Eigenschaft*

$$\int_D |\Delta u|^2\, dx\, dy < +\infty$$

hat, dann ist $w = \Delta u - \varkappa^2 u$ die Lösung der Funktionalgleichung

$$\langle w, \varphi \rangle = \int_{\partial D} h\varphi\, ds , \tag{6.11}$$

wobei φ eine beliebige Funktion aus $\mathcal{H}(D)$ ist.

Beweis: Es sei v die Funktion, die in der Randwertaufgabe (6.6) dem φ entspricht. Es gilt

$$\langle w, \varphi \rangle = \int_D (\Delta u\, \Delta v + \varkappa^2\, \mathrm{grad}\, u\, \mathrm{grad}\, v)\, dx\, dy = \int_D \Delta u\, \varphi\, dx\, dy = \int_{\partial D} \frac{\partial u}{\partial n} \varphi\, ds = \int_{\partial D} h\varphi\, ds ,$$

was zu beweisen war.

Wir betrachten jetzt die Existenz und Eindeutigkeit der Lösung der Funktionalgleichung (6.11). Es sei $A \in \mathcal{C}^4(D) \cap \mathcal{C}^1(D + \partial D)$ die Lösung der Fundamentalrandwertaufgabe für die biharmonische Gleichung

$$\Delta\Delta A = 0 , \qquad A|_{\partial D} = 0 , \qquad \left.\frac{\partial A}{\partial n}\right|_{\partial D} = 0 . \tag{6.12), (6.13), (6.14}$$

Theorem: *Wenn die Lösung der Randwertaufgabe* (6.12-14) *so beschaffen ist, daß*

$$\int_D |\Delta A|^2\, dx\, dy < +\infty \tag{6.15}$$

gilt, dann besitzt die Funktionalgleichung (6.11) *eine eindeutige Lösung in $\overline{\mathcal{H}}(D)$.*

Beweis: Die Lösung von (6.12-14) genügt der Funktionalgleichung

$$\int_D \Delta A\varphi \, dx \, dy = \int_{\partial D} h\varphi \, ds$$

für alle $\varphi \in \mathscr{H}(D)$. Damit wird die Funktionalgleichung (6.11)

$$\langle w, \varphi\rangle = \int_D \Delta A \, \varphi \, dx \, dy \, .$$

Aufgrund von (6.8) folgt, daß das Funktional

$$\mathscr{F}(\varphi) = \int_D \Delta A\varphi \, dx \, dy$$

in $\overline{\mathscr{H}}(D)$ beschränkt ist. Mit der Existenz und Eindeutigkeit folgt nach dem Riesz-Fréchetschen Theorem, daß

$$||w||_{\mathscr{H}} = O(\varkappa) \, ||\Delta A||_{L_2} \, ,$$

demzufolge

$$||w||_{L_2} = O(\varkappa^2) \, ||\Delta A||_{L_2} \, .$$

Anmerkung: In der Entwicklung der oben erwähnten Theorie war die Funktionalgleichung (6.11) wichtig. Sie war zu der Fundamentalrandwertaufgabe (6.1-3) assoziiert. Solche Zuordnung zwischen einer Randwertaufgabe und einer Funktionalgleichung im Hilbertraume wurde zuerst von Rafalson [7] und Greenberg [8], [9] bestimmt und für die numerische Lösung der biharmonischen Gleichung benutzt. In [10], [11] und [12] wurde diese Methode erweitert auf mehrfach zusammenhängende Gebiete und auch auf eine Verallgemeinerung der biharmonischen Gleichung. Im Zusammenhang mit der biharmonischen Gleichung wurde auch eine gleichnamige Methode von Glowinski und Pironneau in einer Reihe von Veröffentlichungen betrachtet ([13] steht für die erste und [14] für eine der letzten Arbeiten aus dieser Reihe).

7. Das asymptotische Verhalten der Lösung der homogenen Gleichung $\Delta\Delta u - \varkappa^2 \Delta u = 0$

Wir betrachten die Integraldarstellung (6.5), wobei w die Lösung der Funktionalgleichung (6.11) ist.

Theorem: *Unter der Bedingung* (6.15) *gilt für die Lösung u der Randwertaufgabe die folgende asymptotische Abschätzung*

$$u(P) = -\frac{1}{\varkappa^2} w + O\left(\frac{\varkappa^2}{\sqrt{\varkappa d}} e^{-\varkappa d}\right), \tag{7.1}$$

wobei $d = \min_{Q \in \partial D} r_{PQ}$ *ist.*

Beweis: Aufgrund der Ergebnisse aus § 4 folgt

$$\frac{1}{2\pi} \int_{\partial D} g(s) \, K_0(\varkappa r_{PQ}) \, ds_Q = O\left(\frac{1}{\sqrt{\varkappa d}} e^{-\varkappa d}\right). \tag{7.2}$$

Betrachten wir jetzt das erste Integral aus (6.5)

$$\frac{1}{2\pi} \int_D w(Q) \, K_0(\varkappa r_{PQ}) \, dx \, dy = \frac{1}{2\pi} \int_{D - D_d(P)} w(Q) \, K_0(\varkappa r_{PQ}) \, dx \, dy + \frac{1}{2\pi} \int_{D_d(P)} w(Q) \, K_0(\varkappa r_{PQ}) \, dx \, dy \, ,$$

wobei $D_d(P)$ der Kreis mit dem Radius d um den Mittelpunkt P ist. Für das erste Integral gilt

$$\frac{1}{2\pi} \left| \int_{D - D_d(P)} w(Q) \, K_0(\varkappa r_{PQ}) \, dx \, dy \right| \leqq \frac{1}{2\pi} K_0(\varkappa d) \int_{D - D_d(P)} |w| \, dx \, dy \leqq \frac{1}{2\pi} K_0(\varkappa d) \int_D |w| \, dx \, dy \leqq$$

$$\leqq \frac{1}{2\pi} K_0(\varkappa d) \, ||w||_{L_2} \sqrt{m(D)} = \frac{k_1}{2\pi} \varkappa^2 K_0(\varkappa d) \, ||\Delta A||_{L_2} \sqrt{m(D)} = O\left(\frac{\varkappa^2}{\sqrt{\varkappa d}} e^{-\varkappa d}\right), \tag{7.3}$$

wobei $m(D)$ der Flächeninhalt von D ist. Aufgrund des Mittelwertsatzes für harmonische Funktionen ergibt sich für das zweite Integral (vgl. auch mit § 4)

$$\frac{1}{2\pi} \int_{D_d(P)} w(Q) \, K_0(\varkappa r_{PQ}) \, dx \, dy = w(P) \int_0^d \varrho K_0(\varkappa \varrho) \, d\varrho = \frac{w(P)}{\varkappa^2} \int_0^{\varkappa d} r K_0(r) \, dr =$$

$$= \frac{w(P)}{\varkappa^2} + \frac{w(P)}{\varkappa^2} O\left(\sqrt{\varkappa d} \, e^{-\varkappa d}\right),$$

oder, weil $|w(P)| \leqq \frac{1}{\sqrt{\pi}\, d} ||w||_{L_2}$ gilt,

$$\frac{1}{2\pi} \int\limits_{D_d(P)} w(Q)\, K_0(\varkappa r_{PQ})\, dx\, dy = \frac{w(P)}{\varkappa^2} + \frac{||w||_{L_2}}{\sqrt{\pi}\, \varkappa^2 d} O\left(\sqrt{\varkappa d}\, e^{-\varkappa d}\right) \frac{w(P)}{\varkappa^2} + ||\Delta A||_{L_2} O\left(\sqrt{\frac{\varkappa}{d}}\, e^{-\varkappa d}\right) \tag{7.4}$$

mit A definiert durch (6.12-14). Aus (7.2-4) folgt (7.1).

Anmerkung: Falls die erste Randbedingung nicht identisch Null ist, dann gilt

$$u(P) = u^*(P) - \frac{1}{\varkappa^2} w(P) + O\left(\frac{\varkappa^2}{\sqrt{\varkappa d}}\, e^{-\varkappa d}\right),$$

wobei u^* die harmonische Funktion ist, die $u^*|_{\partial D} = g$ genügt. Diese Abschätzung entspricht dem Johnschen Ergebnis [1], welches von der Form ist:

$$u(P) = \text{harmonische Funktion} + O(e^{-\varkappa d/2}) .$$

Überdies präzisiert das letzte Theorem die harmonischen Funktionen, für die die oben erwähnte Abschätzung gilt

8. Die inhomogene Gleichung $\Delta\Delta u - \varkappa^2 \Delta u = f$

Wir betrachten die Randwertaufgabe: Man bestimme $u \in \mathscr{C}^4(D) \cap \mathscr{C}^1(D + \partial D)$ als Lösung der Gleichung

$$\Delta\Delta u - \varkappa^2 \Delta u = \varkappa^2 f , \tag{8.1}$$

wobei $f \in \mathscr{C}^0(D + \partial D)$ eine in D definierte gegebene Funktion ist, so daß auf dem Rand

$$u|_{\partial D} = \left.\frac{\partial u}{\partial n}\right|_{\partial D} = 0 . \tag{8.2}$$

Diese Randwertaufgabe ist äquivalent mit: Man bestimme $u \in \mathscr{C}^4(D) \cap \mathscr{C}^1(D + \partial D)$ so, daß

$$\Delta u - \varkappa^2 u = \varkappa^2 F + w , \tag{8.3}$$

$$u|_{\partial D} = \left.\frac{\partial u}{\partial n}\right|_{\partial D} = 0 , \tag{8.4}$$

wobei

$$\Delta F = f , \qquad F|_{\partial D} = 0 , \tag{8.5}$$

und w eine in D harmonische Funktion ist.

Theorem: *Es sei $\varphi \in \mathscr{H}(D)$ und v die Funktion, die in der Randwertaufgabe* (6.6) *dem φ entspricht. Dann gilt für beliebige* $u \in \mathscr{C}^2(D) \cap \mathscr{C}'(D + \partial D)$ so daß

$$u|_{\partial D} = \left.\frac{\partial u}{\partial n}\right|_{\partial D} = 0 , \tag{8.6}$$

die Identität

$$\int\limits_D (\Delta u - \varkappa^2 u)\, \Delta v\, dx\, dy = 0 . \tag{8.7}$$

Beweis: Zum Beweis bemerken wir, daß

$$\Delta\Delta v - \varkappa^2 \Delta v = \Delta\varphi = 0 .$$

Somit ergibt sich aufgrund der Greenschen Formel

$$\int\limits_D \Delta u\, \Delta v\, dx\, dy = \int\limits_D u \Delta\Delta v\, dx\, dy + \int\limits_{\partial D} \left(\frac{\partial u}{\partial n} \Delta v - u \frac{\partial \Delta v}{\partial n}\right) ds = \varkappa^2 \int\limits_D u \Delta v\, dx\, dy ,$$

womit (8.7) bewiesen ist.

Mit Hilfe dieses Theorems und unter der Bedingung, daß $\int\limits_D |\Delta u|^2\, dx\, dy < +\infty$ gilt, ist die Funktion w aus (8.3) bestimmt durch die Funktionalgleichung

$$\langle w, \varphi \rangle = -\varkappa^2 \int F \Delta v\, dx\, dy \tag{8.8}$$

oder, äquivalent,

$$\langle w, \varphi \rangle = \varkappa^2 \int\limits_D \operatorname{grad} F \operatorname{grad} v\, dx\, dy .$$

Nehmen wir $\varphi = w$ an, so folgt

$$||w||^2_{\mathscr{H}} \leqq \varkappa^2 ||\operatorname{grad} F||_{L_2} ||\operatorname{grad} u||_{L_2} \leqq \varkappa ||\operatorname{grad} F||_{L_2} ||w||_{\mathscr{H}} .$$

Man erhält

$$||w||_{\mathscr{H}} \leqq \varkappa ||\operatorname{grad} F||_{L_2} \leqq k_2 \varkappa ||f||_{L_2} , \tag{8.9}$$

wobei k_2 eine Konstante ist, die nur auf dem Gebiet D existiert. Aufgrund von (6.9) folgt

$$\|w\|_{L_2} \leq O(\varkappa^2) \, \|f\|_{L_2} .$$

Anmerkung: Diese letzte Ungleichung kann eine Verbesserung sein, wenn man die asymptotischen Eigenschaften der Lösungen der Randwertaufgabe (6.6) benutzt. Aber für die nächsten Ergebnisse sind solche Verbesserungen überflüssig.

Mit w, bestimmt aus (8.8), wird die Lösung von (3.5)

$$u(P) = -\frac{\varkappa^2}{2\pi} \int_D F(Q) \, K_0(\varkappa r_{PQ}) \, dx \, dy - \frac{1}{2\pi} \int_D w(Q) \, K_0(\varkappa r_{PQ}) \, dx \, dy .$$

Theorem: *Die Lösung der Randwertaufgabe* (8.1-2) *genügt den asymptotischen Beziehungen*

$$u(P) = -\frac{\varkappa^2}{2\pi} \int_D F(Q) \, K_0(\varkappa r_{PQ}) \, dx \, dy - \frac{1}{\varkappa^2} w(P) + O\left(\frac{\varkappa^2}{\sqrt{\varkappa d}} e^{-\varkappa d}\right),$$

$$\frac{\partial u}{\partial x}(P) = -\frac{\varkappa^2}{2\pi} \int_D F(Q) \frac{\partial K_0(\varkappa r_{PQ})}{\partial x} \, dx \, dy - \frac{1}{\varkappa^2} \frac{\partial w}{\partial x}(P) + O\left(\frac{\varkappa^2}{\sqrt{\varkappa d}} e^{-\varkappa d}\right),$$

und

$$\frac{\partial u}{\partial y}(P) = -\frac{\varkappa^2}{2\pi} \int_D F(Q) \frac{\partial K_0(\varkappa r_{PQ})}{\partial y} \, dx \, dy - \frac{1}{\varkappa^2} \frac{\partial w}{\partial x}(P) + O\left(\frac{\varkappa^2}{\sqrt{\varkappa d}} e^{-\varkappa d}\right).$$

Der Beweis dieses Theorems ist entsprechend dem Beweis des Theorems von § 7.

Anmerkung: Aus dem letzten Theorem folgt: Werden in den Lösungen der inhomogenen Gleichungen (8.1) mit homogenen Randbedingungen (8.2) die Glieder von der Ordnung $O\left(\frac{\varkappa^2}{\sqrt{\varkappa d}} e^{-\varkappa d}\right)$ vernachlässigt, dann ist das Ergebnis durch die Lösung der Randwertaufgabe (8.5) und eine bestimmte harmonische Funktion darstellbar. Benützen wir (4.9) und (8.9), so erhalten wir im Innern von D

$$u(P) = -F(P) - \frac{w(P)}{\varkappa^2} + O\left(\frac{1}{\varkappa^2}\right)$$

und die entsprechende Formel für die Ableitung erster Ordnung. Diese Formeln entsprechen in bezug auf die Punktkonvergenz denen von Friedman [2], Greenlee [3], Lions [4] und Schuss [5], was die L_2-Konvergenz betrifft.

Literatur

1 John, F., The transition from thin plate to membrane in the case of a plate under uniform tension, in: Continuum Mechanics and Related Problems of Analysis, Moscow 1972.
2 Friedmann, A., Partial Differential Equations, Holt, Reinhart and Wiston, New York 1969.
3 Greenlee, W. M., Rate of convergence in singular perturbations, Ann. Inst. Fourier (Grenoble) **18** 2, 135–191 (1969).
4 Lions, J. L., Perturbations singulières dans les problèmes aux limites et en controle optimal, Lecture Notes in Math., Springer-Verlag, Berlin 1973.
5 Schuss, Z., Singular perturbations and the transition from thin plate to membrane, Proc. Amer. Math. Soc. **38**, 139–147 (1976).
6 Tychonoff, A. M.; Samarski, A. A., Differentialgleichungen der Mathematischen Physik, VEB Deutscher Verlag der Wissenschaften, Berlin 1959.
7 Рафалсон, З. Л., К вопросу о решении бигармонического уравнения, Доклады Акад. Наук СССР **64**, 799—802 (1949).
8 Гринберг, Г. А., О решении плоской задачи теории упругости и задачи об изгибе тонкой плиты с закрепленным контуром, Доклады Акад. Наук СССР **76**, 661—664 (1951).
9 Гринберг, Г. А.; Лебедев, Н. Н.; Уфлянд, Я. С., Метод решения общей бигармонической задачи для прямоугольной области заданной наконтурах значении функции и ее нормальной производной, Приклад. Мат. Мех. **27**, 73—86 (1953).
10 Mazilu, P., Metoda de rezolvare a unei probleme plane a elasticitatii (rum.), Studii Cerc. Mat. **20**, (525–543 (1968).
11 Mazilu, P., Aplicatii ale teoriei functiilor complexe in teoria elasticitatii meaiilor neomogene (rum.). Studii Cerc. Mat. **23**, 1219–1315 (1971).
12 Mazilu, P., Sur un problème plan de la théorie de l'élasticité des milieux hétérogènes. C. R. Acad. Sc. Paris, 268, 1969.
13 Glowinski, R.; Pironneau, O., On solving a mixed finite element approximation of the Dirichlet problem for the biharmonic operator by a "quasi-direct" method and various iterative methods, in "Mathematical Aspects of Finite Element Methods", Lecture Notes in Mathematics 606, Springer-Verlag, Berlin **1975**, 167–203.
14 Glowinski, R.; Pironneau O., Numerical methods for the first biharmonic equation and for the two-dimensional Stokes problem, SIAM Rev. **21**, 167–212 (1979).
15 Berg, L., Asymptotische Darstellungen und Entwicklungen, Berlin 1968.

Eingereicht: 31. 10. 1980, revidierte Fassung: 6. 2. 1981

Anschrift: Dr. P. Mazilu, Ruhr-Universität Bochum, Lehrstuhl für Mechanik I, Postfach 102148, D-4630 Bochum 1, BRD

ZAMM 62, 27 – 33 (1982)

J. HASLINGER / J. LOVÍŠEK

The Obstacle Problem for the Equilibrium of a Shallow Shell Reinforced with Stiffening Ribs

Die Arbeit behandelt die Existenz und Eindeutigkeit der Lösung einer Variationsungleichung für das Hindernisproblem bei einer mit Versteifungsrippen verstärkten flachen Schale. Das Hindernisproblem wird als direktes Variationsproblem formuliert. Danach wird eine Finite-Elemente-Approximation für das Hindernisproblem konstruiert und eine Fehlerabschätzung für die Approximation bewiesen.

The paper deals with the existence and uniqueness of a solution of a variational inequality for the obstacle problem of the shallow shell reinforced with stiffening ribs. The obstacle problem is formulated as a direct variational problem. Next we construct a finite element approximation of the obstacle problem, and we prove an error estimate of the approximation.

Статья посвящена существованием и единственностью решения некоторого вариационного неравенства для задачи препятствия пологой оболочки усиливанной ребрами жесткости. Задача препятства формулируется как прямая вариационная задача. Потом строится приближение задачи препятства с помощью метода конечных элементов и доказывается оценка ошибок для приближения.

Let us suppose that the deflection w of the shell is limited from below by a rigid obstacle. Mathematically it means that w is greater or equal to a prescribed function, describing the obstacle. The obstacle problem consists in finding the equilibrium position of a shallow shell reinforced with stiffening ribs, and is formulated as a variational inequality of elliptic type. The variational formulation is obtained from the principle of virtual works, using admissible virtual displacements. In order to prove the existence of a solution this problem leads to a variational inequality over a certain convex set. We use a finite element approximation of the shell problem. The proof of convergence for Bogner-Fox-Schmidt rectangle elements is given.

1. Formulation of the problem

A *shallow shell* (Fig. 1) is a thin planar construction the tension of which is small compared with the radius of curvature of the middle surface. The theory of shallow shells is based on the assumption that the middle surface of the shell with its small tension obeys the laws of Euclidean geometry in the plane with greatest values of projection of the whole surface. We suppose that the middle surface of the shell can be expressed in Cartesian coordinates (x, y, z) by the equation $z = z(x, y)$. Then under the assumptions $[\partial z / \partial x| \ll 1; |\partial z | \partial y| \ll 1$ we have

$$\frac{1}{R_x} \approx -\frac{\partial^2 z}{\partial x^2}; \quad \frac{1}{R_y} \approx -\frac{\partial^2 z}{\partial y^2}; \quad \frac{1}{R_{xy}} \approx -\frac{\partial^2 z}{\partial x\, \partial y}, \tag{1.1}$$

where R_x, R_y and R_{xy} are the normal radii of curvature and the twist of the surface for the (x, y)-coordinate curves, respectively.

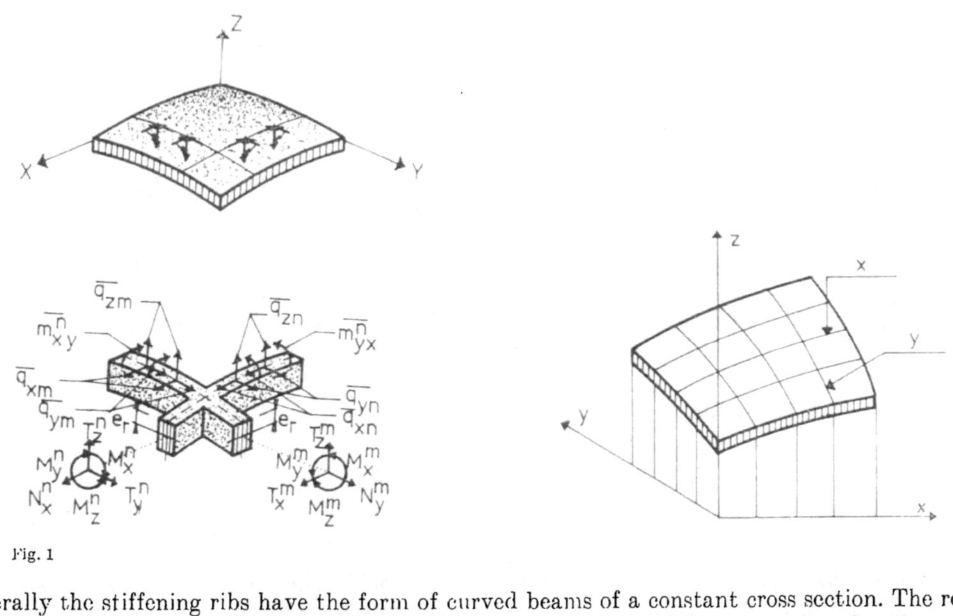

Fig. 1

Generally the stiffening ribs have the form of curved beams of a constant cross section. The reference axis of the rib is the line connecting the centres of all cross sections of the beam. We assume that the ribs are placed on the inner side of the shell ($e_r > 0$, Fig. 1). We denote by

$$q_n = \langle q_{xn}, q_{yn}, q_{zn}, m_{yx}\rangle^T, \quad q_m = \langle q_{xm}, q_{ym}, q_{zm}, m_{xy}\rangle^T \tag{1.2}$$

the vectors of the unknown contact forces on the boundary of the contact between the shell and the n-th rib (in the direction X) and the m-th rib (in the direction Y), respectively (positively directed stress resultants acting on one shell element, the displacement vector components, and external loading force components illustrated by figure 1). The equilibrium equation for the shell in operator form is

$$[P]\,\boldsymbol{u} = \boldsymbol{p} \tag{1.3}$$

where $\boldsymbol{p} = \langle X,\ Y,\ Z\rangle^{T} = \langle R_x/B,\ R_y/B,\ R_z + p/D\rangle^{T}$ is the vector of the surface strains (known and unknown the exterior forces)

$$\begin{aligned}
R_x &= \sum_{i=1}^{n-1} (p_{x_i} - \bar{q}_{x_i})\,\delta(x - x_i) - \sum_{j=1}^{m-1} \bar{q}_{x_j}\delta(y - y_j) + \sum_{i=1}^{n-1}\sum_{j=1}^{m-1} F_{xij}\delta(x - x_i)\,\delta(y - y_j)\,,\\
R_y &= \sum_{j=1}^{m-1} (p_{y_j} - \bar{q}_{y_j})\,\delta(y - y_j) - \sum_{i=1}^{n-1} \bar{q}_{yi}\delta(x - x_i) + \sum_{i=1}^{n-1}\sum_{j=1}^{m-1} F_{xij}\delta(x - x_i)\,\delta(y - y_j)\,,\\
R_z &= \sum_{i=1}^{n-1} (p_{z_i} - \bar{q}_{z_i})\,\delta(x - x_i) + \sum_{j=1}^{m-1} (p_{z_j} - \bar{q}_{z_j})\,\delta(y - y_j) + \sum_{i=1}^{n-1}\sum_{j=1}^{m-1} F_{zij}\delta(x - x_i)\,\delta(y - y_j)\,,\\
M_x &= -\sum_{i=1}^{n-1} \bar{m}_{xi}\frac{\partial\delta}{\partial y}(x - x_i); \qquad M_y = -\sum_{j=1}^{m-1} \bar{m}_{yj}\frac{\partial\delta}{\partial x}(y - y_j)\,,
\end{aligned} \tag{1.4}$$

$\delta(\ )$ is the Dirac measure; $p_{x_i}, p_{y_j}, p_{z_i}, p_{z_j}$ are the densities per unit of length of exterior straight lined load on the lines $x = x_i$, $y = y_j$; $F_{xij}, F_{yij}, F_{zij}$ the projections of a singular force acting in the point (x_i, y_j) to the axes $X,\ Y,\ Z$; $p(x, y)$ is the density per unit surface of the vertical exterior forces. $\bar{q}_{x_i}, \bar{q}_{x_j}, \bar{q}_{y_i}, \bar{q}_{y_j}, \bar{q}_{z_i}, \bar{q}_{z_j}, \bar{M}_x, \bar{M}_y$ are the elements of unknown general specific load in the Kirchhoff sense acting on the contact boundary with the i-th transverse rib and the j-th longitudinal rib. (Fig. 2). $D = \dfrac{Eh^3}{12(1 - \mu^2)}$ denotes cylindrical stiffness; $B = \dfrac{Eh}{1 - \mu^2}$ stiffness of tension and $0 < \mu < 1/2$ Poisson's ratio; $\boldsymbol{u} = \langle u,\ v,\ w\rangle^{T}$ is the displacement vector of the middle surface of the shell.

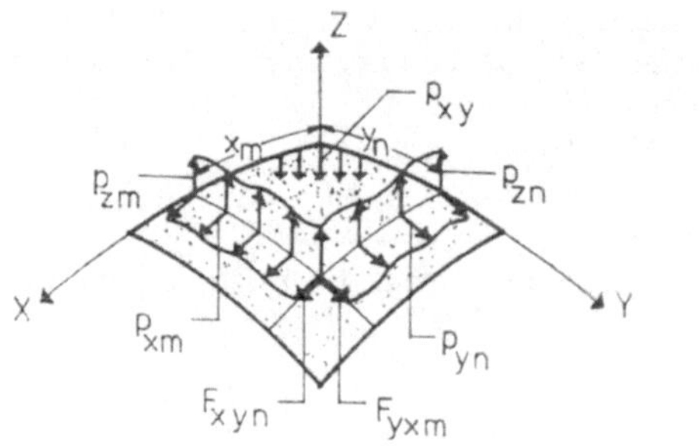

Fig. 2.

The elements of the matrix $[P]$ are differential operators exactly expressed in the form

$$[P] = \begin{bmatrix}
\dfrac{\partial^2}{\partial x^2} + \dfrac{1-\mu}{2}\dfrac{\partial^2}{\partial y^2}, & \dfrac{1+\mu}{2}\dfrac{\partial^2}{\partial x\,\partial y}, & 0\\[2ex]
\dfrac{1+\mu}{2}\dfrac{\partial^2}{\partial x\,\partial y}, & \dfrac{1-\mu}{2}\dfrac{\partial^2}{\partial x^2} + \dfrac{\partial^2}{\partial y^2}, & 0\\[2ex]
0, & 0, & \dfrac{D}{B}\nabla^2\nabla^2
\end{bmatrix} + \begin{bmatrix}
0, & 0, & -(k_x + \mu k_y)\dfrac{\partial}{\partial x}\\[2ex]
0, & 0, & -(k_x + \mu k_y)\dfrac{\partial}{\partial y}\\[2ex]
(k_x + \mu k_y)\dfrac{\partial}{\partial x}, & (k_y + \mu k_x)\dfrac{\partial}{\partial y}, & -(k_x^2 + k_y^2 + 2\mu k_x k_y)
\end{bmatrix} \tag{1,5}$$

The equations of statical compatibility for thin curved beams are expressed in the form

$$[R]\,\boldsymbol{u}_r = \boldsymbol{q}_r$$

with

$$\boldsymbol{u}_r = \langle u_r,\ v_r,\ w_r,\ \theta_r\rangle^{T},$$

where u_r, v_r, w_r, express the displacements of the points of the axes of the rib and θ_r expresses the rotation of the cross section of the rib. Next we have $\boldsymbol{q}_r$ being the vector of generalized forces, and the elements of the matrix $[R]$ being differential operators we introduce below. We suppose that the contact boundary between the rib and the shell is only on the intersection of the perpendicular plane of the rib and the middle suface of the shell. The perpendicular axis Z of the rib is the main central inertia axis, which is orthogonal to the supporting surface (R_x, R_y) of the shell. We therefore formulate the contact conditions in the following form (Fig. 3).

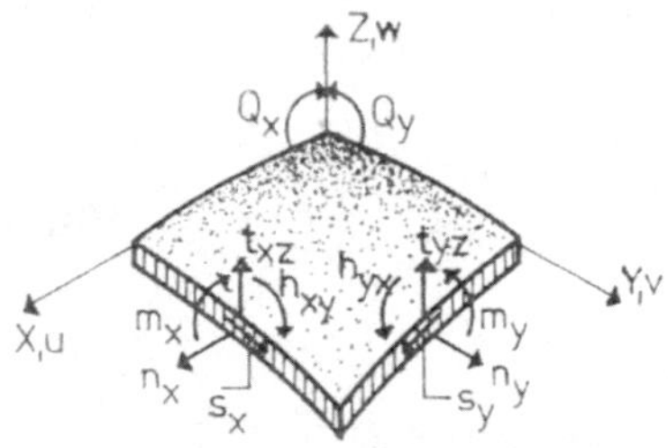

Fig. 3.

1° statical conditions:

$$\bar{q}_{xn} = -\Delta n_x; \qquad q_{yn} = -\Delta s_x; \qquad q_{zn} = \Delta t_{xz}; \qquad \bar{m}_{yx} = -\Delta m_y,$$

2° geometrical conditions:

$$u_r^* = u; \qquad v_r^* = v \qquad w_r^* = w,$$

$\theta_r = \theta_x$ (rotation of the orthogonal element of the shell)

where $u_r^* = u_r - e_r\theta_r; \qquad v_r^* = v_r - e_r \dfrac{\mathrm{d}w_r}{\mathrm{d}y}$.

It follows from relation (1.6) that the unknown vector of the load which acts on the contact surface between the shell and the n-th and the m-th rib respectively has the form ([9], [11])

$$\boldsymbol{q}_n = [R_n]\boldsymbol{U}; \qquad \boldsymbol{q}_m = [R_m]\boldsymbol{U}, \tag{1.8}$$

where

$$\boldsymbol{U} = (\boldsymbol{u}, \theta_x),$$

$$[R_n] = \begin{bmatrix} b_{11} & 0 & 0 & b_{14} \\ 0 & b_{22} & b_{23} & 0 \\ 0 & b_{32} & b_{33} & 0 \\ b_{41} & 0 & 0 & b_{44} \end{bmatrix}$$

and

$$b_{11} = D_{rx}\frac{\mathrm{d}^4}{\mathrm{d}y^4}; \qquad b_{22} = -B_r\frac{\mathrm{d}^2}{\mathrm{d}y^2};$$

$$b_{33} = D_{rx}\left[\frac{\mathrm{d}^4}{\mathrm{d}y^4} + \frac{1}{\varrho_n^2}\left(e_r\frac{\mathrm{d}^2}{\mathrm{d}y^2} + k_y\right)^2\right],$$

$$b_{44} = -\left(G_r + k_y D_{rx} e_r - D_{rx} e_r^2 \frac{\mathrm{d}^2}{\mathrm{d}y^2}\right)\frac{\mathrm{d}^2}{\mathrm{d}y^2},$$

$$b_{14} = -b_{41} = D_{rx}\left(e_r\frac{\mathrm{d}^2}{\mathrm{d}y^2} - k_y^2\right)\frac{\mathrm{d}^2}{\mathrm{d}y^2} - k_y G_r\frac{\mathrm{d}^2}{\mathrm{d}y^2}, \quad b_{23} = -b_{32} = B_r\left[(k_y\varrho_r^2 - e_r)\frac{\mathrm{d}^2}{\mathrm{d}y^2} - k_y\frac{\mathrm{d}}{\mathrm{d}y}\right],$$

D_{rx}, B_r, G_r are the flexural, bending and torsional rigidities of the rib, respectively, ϱ_r is the cross section function (the radius of inertia of the cross section surface of the rib), k_y, k_x are the main curvatures (constant) of the shell, e_r is the distance between the neutral axis of the cross section surface of the rib and the inner surface of the rib. The matrix $[R_m]$ is of the same form with the operator $\dfrac{\mathrm{d}}{\mathrm{d}x}$ instead of $\dfrac{\mathrm{d}}{\mathrm{d}y}$.

2. Statement of the problem

Let $\Omega = (a, b) \times (c, d)$, $a < b$, $c < d$ be the projection of the middle surface of the shell and let I_i, I_j be the projections of stiffening ribs, given by

$$I_i = \{x_i\} \times (c, d), \qquad i = 1, \dots, n, \qquad I_j = (a, b) \times \{y_j\}, \qquad j = 1, \dots, m,$$

where

$$a < x_1 < x_2 < \dots < x_n < b, \qquad c < y_1 < y_2 < \dots < y_m < d.$$

By Γ we denote the system of I_i, I_j for all i, j.

Let $\mathcal{D}(\Omega)$ be the set of all infinitely many differentiable functions with a compact support in Ω. By $H^k(\Omega)$ ($k \geq 0$, integer) we denote the classical Sobolev space of functions the derivatives of which up to order k are from

$H^0(\Omega) \equiv L^2(\Omega)$, i.e. square integrable. $H^k(\Omega)$ is a Hilbert space with the norm

$$\|u\|_k = \Big(\int_\Omega \sum_{|\alpha| \leq k} |D^\alpha u|^2 \, d\Omega\Big)^{1/2}, \tag{2.1}$$

where $D^\alpha u = \partial^{|\alpha|} u / \partial x^{\alpha_1} \partial y^{\alpha_2}$, $\alpha = (\alpha_1, \alpha_2)$, α_i are non-negative integers, $|\alpha| = \alpha_1 + \alpha_2$ and $D^0 u = u$. $H_0^k(\Omega)$ is defined as the closure of $\mathcal{D}(\Omega)$ in $H^k(\Omega)$-norm.

Let V be the space defined by

$$V = H_0^1(\Omega) \times H_0^1(\Omega) \times H_0^2(\Omega) .$$

In order to give the variational formulation of the problem of a clamped shell reinforced with stiffening ribs, we introduce the space W by

$$\begin{aligned} W = \{z = (u, v, w) \in V \mid\ & u|_j \in H_0^1(a, b), \quad u|_i \in H_0^2(c, d), \quad u|_j(x_i) = u|_i(x_j); \quad v|_j \in H_0^2(a, b), \\ & v|_i \in H_0^1(c, d), \quad v|_j(x_i) = v|_i(x_j); \ w|_j \in H_0^2(a, b), \quad w|_i \in H_0^2(c, d), \\ & \partial w/\partial y|_j \in H_0^2(a, b), \quad \partial w/\partial x|_i \in H_0^2(c, d), \quad \partial w/\partial x|_i (y_j) = \partial w|_j/\partial x(x_i), \\ & \partial w/\partial y|_j(x_i) = \partial w|_i/\partial y(y_j), \quad \partial^2 w/\partial x\, \partial y|_j(x_i) = \partial^2 w/\partial y\, \partial x|_i(y_j)\}, \end{aligned}$$

where $\theta|_i$ and $\theta|_j$ are the traces of θ on I_i and I_j, respectively.

Now, let us set

$$(z_1, z_2)_W = (z_1, z_2)_V + (z_1, z_2)_\Gamma, \quad z_j = (u_j, v_j, w_j) \in W, \quad j = 1, 2, \tag{2.2}$$

where

$$(z_1, z_2)_V = (u_1, u_2)_1 + (v_1, v_2)_1 + (w_1, w_2)_2$$

and

$$\begin{aligned} (z_1, z_2)_\Gamma = & \sum_{j=1}^{m} \int_a^b \left\{ \frac{\partial u_1|_j}{\partial x} \frac{\partial u_2|_j}{\partial x} + \frac{\partial^2 v_1|_j}{\partial x^2} \frac{\partial^2 v_2|_j}{\partial x^2} + \frac{\partial^2 w_1|_j}{\partial x^2} \frac{\partial^2 w_2|_j}{\partial x^2} + \frac{\partial^2}{\partial x^2}\left(\frac{\partial w_1}{\partial y}\Big|_j\right) \cdot \frac{\partial^2}{\partial x^2}\left(\frac{\partial w_2}{\partial y}\Big|_j\right) \right\} dx + \\ & + \sum_{i=1}^{n} \int_c^d \left\{ \frac{\partial^2 u_1|_i}{\partial y^2} \frac{\partial^2 u_2|_i}{\partial y^2} + \frac{\partial v_1|_i}{\partial y} \frac{\partial v_2|_i}{\partial y} + \frac{\partial^2 w_1|_i}{\partial y^2} \frac{\partial^2 w_2|_i}{\partial y^2} + \frac{\partial^2}{\partial y^2}\left(\frac{\partial w_1}{\partial x}\Big|_i\right) \frac{\partial^2}{\partial y^2}\left(\frac{\partial w_2}{\partial x}\Big|_i\right) \right\} dy . \end{aligned}$$

It is easy to see that $(\ ,\ )_W$ is a scalar product in W. The corresponding norm will be denoted by $\|\cdot\|_W$.

Theorem 2.1: *The set W with the scalar product* (2.2) *is a Hilbert space.*

The proof is given in [2].

The following density result is very important to obtain the convergence of finite element approximations, which will be studied in Section 4.

Theorem 2.2: *The set $[\mathcal{D}(\Omega)]^3$ is dense in W with respect to the W-norm.*

The proof can be found in [2].

3. Variational formulation of the obstacle problem

First we introduce some notations useful in what follows. Let

$$\{\varepsilon(z)\}^T = (\varepsilon_x, \varepsilon_y, \gamma_{xy}), \qquad \{\varkappa(z)\}^T = (\varkappa_x, \varkappa_y, \varkappa_{xy}),$$

where

$$\varepsilon_x \equiv \varepsilon_x(z) = \frac{\partial u}{\partial x} + k_x w, \quad \varepsilon_y = \varepsilon_y(z) = \frac{\partial v}{\partial y} + k_y w, \quad \gamma_{xy} = \gamma_{xy}(z) = \frac{\partial u}{\partial y} + \frac{\partial v}{\partial x};$$

$$\varkappa_x = \varkappa_x(z) = \frac{\partial^2 w}{\partial x^2}, \quad \varkappa_y = \varkappa_y(z) = \frac{\partial^2 w}{\partial y^2}, \quad \varkappa_{xy} = \varkappa_{xy}(z) = 2\frac{\partial^2 w}{\partial x\, \partial y}, \qquad z = (u, v, w) \in W .$$

The *stiffness matrix* is defined by

$$[D] = \frac{E}{1-\mu^2} \begin{bmatrix} 1 & \mu & 0 \\ \mu & 1 & 0 \\ 0 & 0 & (1-\mu)/2 \end{bmatrix}, \qquad 0 < \mu < 1 .$$

We further introduce the following vectors and matrices

$$\{\varepsilon_i(z)\}^T = \left(\frac{\partial u|_i}{\partial y}, \frac{\partial^2 u|_i}{\partial y^2}, \frac{\partial v|_i}{\partial y}, \frac{\partial^2 w|_i}{\partial y^2}, e_i \frac{\partial^2 w|_i}{\partial y^2} + k_y w|_i, \frac{\partial}{\partial y}\left(\frac{\partial w}{\partial x}\Big|_i - k_y u|_i\right), \frac{\partial^2}{\partial y^2}\left(\frac{\partial w}{\partial x}\Big|_i - k_y u|_i\right)\right),$$

$$[K_i] = \begin{bmatrix} 0 & 0 & 0 & 0 & 0 & k_y(D_i + G_i) & 0 \\ 0 & D_i & 0 & 0 & 0 & 0 & D_i e_i \\ 0 & 0 & B_i & -B_i k_y \varrho_i^2 & -B_i & 0 & 0 \\ 0 & 0 & -B_i k_y \varrho_i^2 & D_i & 0 & 0 & 0 \\ 0 & 0 & -B_i & 0 & D_i \varrho_i^{-2} & 0 & 0 \\ -k_y(D_i + G_i) & 0 & 0 & 0 & 0 & G_i + k_y D_i e_i & 0 \\ 0 & -D_i e_i & 0 & 0 & 0 & 0 & D_i e_i^2 \end{bmatrix} \quad \text{for} \quad i = 1, 2, \dots, n,$$

and

$$\{\varepsilon_j(z)\}^\mathsf{T} = \left(\frac{\partial v|_j}{\partial x}, \frac{\partial^2 v|_j}{\partial x^2}, \frac{\partial u|_j}{\partial x}, \frac{\partial^2 w|_j}{\partial x^2}, e_j \frac{\partial^2 w|_j}{\partial x^2} + k_x w|_j, \frac{\partial}{\partial x}\left(\frac{\partial w}{\partial y}\bigg|_j - k_x v|_j\right), \frac{\partial^2}{\partial x^2}\left(\frac{\partial w}{\partial y}\bigg|_j - k_x v|_j\right)\right), \quad z = (u, v, w),$$

$$[K_j] = \begin{bmatrix} 0 & 0 & 0 & 0 & 0 & k_x(D_j + G_j) & 0 \\ 0 & D_j & 0 & 0 & 0 & D_j e_j & 0 \\ 0 & 0 & B_j & -B_j k_x \varrho_j^2 & -B_j & 0 & 0 \\ 0 & 0 & -B_j k_x \varrho_j^2 & D_j & 0 & 0 & 0 \\ 0 & 0 & -B_j & 0 & D_j \varrho_j^{-2} & 0 & 0 \\ -k_x(D_j + G_j) & 0 & 0 & 0 & 0 & G_j + k_x D_j e_j & 0 \\ 0 & -D_j e_j & 0 & 0 & 0 & 0 & D_j e_j^2 \end{bmatrix} \quad \text{for} \quad j = 1, 2, \dots, m.$$

The sense of all symbols has been introduced in Section 1.

In order to give the variational formulation of the problem in question, we introduce the following set of kinematically admissible horizontal and vertical displacements:

$$K = \{z \in W \mid z = (u, v, w),\ w \geqq \psi \text{ a.e. in } \Omega\},$$

where $\psi \in C(\overline{\Omega})$ is a given function, describing the obstacle and such that $\psi \leqq 0$ on $\partial\Omega$. It is readily seen that K is a non-empty, closed and convex subset of W. Now, we introduce the following bilinear forms defined on $W \times W$ as follows:

$$A(z_1, z_2) = h \int_\Omega \{\varepsilon(z_1)\}^\mathsf{T} [D] \{\varepsilon(z_2)\} \,\mathrm{d}\Omega + \frac{h^3}{12} \int_\Omega \{\varkappa(z_1)\}^\mathsf{T} \cdot [D] \{\varkappa(z_2)\} \,\mathrm{d}\Omega,$$

$$b_i(z_1, z_2) = \int_{I_i} \{\varepsilon_i(z_1)\}^\mathsf{T} [K_i] \{\varepsilon_i(z_2)\} \,\mathrm{d}y, \qquad i = 1, \dots, n,$$

$$b_j(z_1, z_2) = \int_{I_j} \{\varepsilon_j(z_1)\}^\mathsf{T} [K_j] \cdot \{\varepsilon_j(z_2)\} \,\mathrm{d}x, \qquad j = 1, \dots, m$$

for all $z_1, z_2 \in W$. Now we set

$$b(z_1, z_2) = \sum_{i=1}^{n} b_i(z_1, z_2) + \sum_{j=1}^{m} b_j(z_1, z_2),$$

$$a(z_1, z_2) = A(z_1, z_2) + b(z_1, z_2).$$

Let $p \in [L^2(\Omega)]^3$. We define a linear functional l by the relation

$$l(z) = \int_\Omega p \cdot z \,\mathrm{d}\Omega + \sum_{i=1}^{n} \sum_{j=1}^{m} \Big\{ u|_i(y_j) + u|_j(x_i) + v|_i(y_j) + v|_j(x_i) + w|_i(y_j) + w|_j(x_i) + $$
$$+ \left(\frac{\partial w}{\partial x}\bigg|_i - k_y u|_i\right)(y_j) + \left(\frac{\partial w}{\partial y}\bigg|_j - k_x v|_j\right)(x_i)\Big\}, \qquad z = (u, v, w) \in W.$$

Clearly l is bounded on W, as follows from the definition of $\|\ \|_W$.

We call a *variational* (weak) *solution* of the obstacle problem for a shallow shell reinforced with stiffening ribs a function $z_0 = (u_0, v_0, w_0) \in K$ such that:

$$a(z_0, z - z_0) \geqq l(z - z_0) \quad \text{for all} \quad z \in K. \tag{$\mathfrak{P}$}$$

Theorem 3.1: *Let $k_x \in C^2(\langle a, b\rangle)$, $k_y \in C^2(\langle c, d\rangle)$, $\varrho_i \ll D_i, B_i, G_i, k_y$ and $\varrho_j \ll D_j, B_j, G_j, k_x$, $i = 1, \dots, n$, $j = 1, \dots, m$. Then there exists a unique solution $z_0 \in K$ of ($\mathcal{P}$).*

Proof: If ϱ_i and ϱ_j are sufficiently small, the bilinear form $a(z_1, z_2)$ is W-elliptic, i.e. there exists a positive constant $\alpha > 0$ such that

$$a(z, z) \geqq \alpha \|z\|_W^2 \quad \text{for all} \quad z \in W. \tag{3.1}$$

The proof of (3.1) is given in [2]. From this and the fact that K is a closed, convex subset of W, the existence and the uniqueness of z_0 follows.

4. Numerical approximation of $(\mathcal{P})$

The approximation of $(\mathcal{P})$ consists of two steps:

(1°) the replacement of $(\mathcal{P})$ by a sequence of "finite-dimensional" problems $(\mathcal{P}_h)$,

(2°) the numerical solution of $(\mathcal{P}_h)$.

To study (1°) we use the finite element method. Let $\{\mathcal{R}_h\}$, $h \to 0+$ be a *regular* family of *rectangulations* of $\overline{\Omega}$, what means that $\overline{\Omega}$ is expressed as the union of finite number of closed rectangles R_i, $i = 1, \dots, n(h)$, with usual requirements on their mutual position. Let h denote the maximal side of all R_i belonging to the same rectangulation. Moreover, we shall suppose that

$$\mathring{R}_i \cap I_j = \emptyset\,, \qquad \mathring{R}_i \cap I_k = \emptyset\,, \qquad i = 1, \dots, n(h)\,, \qquad j = 1, \dots, n\,, \qquad k = 1, \dots, m\,. \tag{4.1}$$

(4.1) means that all ribs I_j, I_k are the sides of R_i, only.

Let $\mathfrak{N}_h$, $\mathring{\mathfrak{N}}_h$ be the set of all nodes and interior nodes, respectively, of the rectangulation $\mathcal{R}_h$. Next, we introduce families of $\{\mathcal{R}_h\}$ satisfying

$$\mathfrak{N}_{h_1} \subset \mathfrak{N}_{h_2} \text{ if } h_1 > h_2\,. \tag{4.2}$$

By $Q_3(R_i)$ we denote the set of all bicubic functions defined on R_i, i.e.

$$q \in Q_3(R_i) \Leftrightarrow q(x, y) = \sum_{\substack{1 \leqq i \leqq 3\\ 1 \leqq j \leqq 3}} \alpha_{ij} x^i y^i\,, \qquad (x, y) \in R_i\,.$$

To every $\mathcal{R}_h$ we associate the finite-dimensional spaces X_h, Y_h defined by

$$X_h = \{v_h \in C^1(\overline{\Omega})|\ v_h|_{R_i} \in Q_3(R_i) \quad \text{for all } R_i \in \mathcal{R}_h,\ v_h = 0 \text{ on } \partial\Omega\} \subset H_0^1(\Omega)\,,$$

$$Y_h = \{v_h \in C^1(\overline{\Omega})|\ v_h|_{R_i} \in Q_3(R_i) \quad \text{for all } R_i \in \mathcal{R}_h,\ v_h = \partial v_h/\partial n = 0 \text{ on } \partial\Omega\} \subset H_0^2(\Omega)\,.$$

It is easy to verify that

$$W_h = X_h \times X_h \times Y_h$$

is a finite-dimensional subspace of W.

Remark 4.1: Every function $v_h \in X_h$, Y_h is uniquely determined by $v_h(a_i)$, $\partial v_h(a_i)/\partial x$, $\partial v_h(a_i)\partial y$ and $\partial^2 v_h(a_i)/\partial x\,\partial y$ at $a_i \in \mathfrak{N}_h$ (see [4]). X_h, Y_h are generated by the so called *Bogner-Fox-Schmidt rectangular element.*

By K_h we denote *the finite-dimensional approximation of* K, defined by

$$K_h = \{z_h = (u_h, v_h, w_h) \in W_h|\ w_h(a_i) \geqq \psi(a_i) \quad \text{for all } a_i \in \mathring{\mathfrak{N}}_h\}\,.$$

The problem of finding $z_h^0 \in K_h$, such that

$$(\mathcal{P}_h) \qquad a(z_h^0, z_h - z_h^0) \geqq l(z_h - z_h^0) \quad \text{for all} \quad z_h \in K_h$$

will be called *the approximation* of $(\mathcal{P})$.

Arguing in the same manner as in Theorem 3.1 we obtain

Theorem 4.1: *Let all assumption of theorem 3.1 be satisfied. Then there exists a unique solution* z_h^0 *of* $(\mathcal{P}_h)$ *for every* $h \in (0, 1)$.

Our aim is to prove the convergence of z_h^0 to z^0 in the W-norm. To this end we assume that

$$\psi < 0 \quad \text{on} \quad \partial\Omega\,. \tag{4.3}$$

First, we prove some auxiliary lemmas, which will be needed in what follows.

Lemma 4.1: *Let* (4.3) *be satisfied. Then to every* $z \in K$ *there exists* $\{z_h\}$, $z_h \in K_h$ *such that*

$$z_h \to z\,, \qquad h \to 0+$$

in W.

Proof: Let $z = (u, v, w) \in K$. If $u, v, w \in H^4(\Omega) \cap H_0^2(\Omega)$, we can choose $z_h = (r_h u, r_h v, r_h w)$, where $r_h\theta|_{R_i}$ is the Q_3-interpolate of θ on R_i. Clearly $z_h \in K_h$ and

$$\|z - z_h\|_W \to 0\,, \qquad h \to 0+\,, \tag{4.4}$$

follows from well-known interpolating properties of X_h, Y_h and the definition of $\|\ \|_W$ (see [4]).

Let $z = (u, v, w) \in K$ be an arbitrary given function. We choose a function $\tilde{\Phi} \equiv (0, 0, \Phi) \in W$ such that

$$\Phi > 0 \quad \text{in } \Omega \quad \text{and} \quad \|\tilde{\Phi}\|_W = 1\,. \tag{4.5}$$

Setting $z_\varepsilon = z + \varepsilon\tilde{\Phi}$, $\varepsilon > 0$, and using (4.3) and (4.5), we obtain

$$\|z_\varepsilon - z\|_W = \varepsilon\,, \qquad z_\varepsilon \geqq z\,, \qquad w_\varepsilon > \psi \quad \text{in } \Omega\,, \tag{4.6}$$

where $w_\varepsilon = w + \varepsilon\Phi$. Hence $z_\varepsilon \in K$ follows. Using the density result of Theorem 2.2, we can find a sequence $\{z_{\varepsilon N}\}$, $z_{\varepsilon N} \in [\mathcal{D}(\Omega)]^3$ such that

$$z_{\varepsilon N} \to z_\varepsilon\,, \qquad N \to +\infty \quad \text{in } W\,. \tag{4.7}$$

From this and the imbedding $H_0^2(\Omega)$ into $C(\bar{\Omega})$, we may write

$$w_{\varepsilon N} \rightrightarrows w\,, \qquad N \to \infty \text{ (uniformly) in } \bar{\Omega}\,.$$

If N is sufficiently large, $w_{\varepsilon N} > \psi$ in $\bar{\Omega}$, so that $z_{\varepsilon N} \in K \cap [\mathcal{D}(\Omega)]^3$. The first part of the proof ensures the existence of $\{z_h\}$, $z_h \in K_h$, such that

$$\|z_{\varepsilon N} - z_h\|_W \to 0+ \quad \text{if} \quad h \to 0+\,. \tag{4.8}$$

From the triangle inequality and (4.6), (4.7), (4.8) follows

$$\|z - z_h\|_W \leqq \|z - z_\varepsilon\|_W + \|z_\varepsilon - z_{\varepsilon N}\|_W + \|z_{\varepsilon N} - z_h\|_W \to 0+ \quad \text{if} \quad \varepsilon\,, h \to 0+\,, \qquad N \to +\infty\,.$$

Lemma 4.2: *Let (4.2) be satisfied. Let $\{z_h\}$, $z_h \in K_h$, be such that*

$$z_h \rightharpoonup z \text{ (weakly) in } W\,. \tag{4.9}$$

Then $z \in K$.

Proof: Let $z = (u, v, w)$, $z_h = (u_h, v_h, w_h)$. It is sufficient to prove that $w \geqq \psi$ a.e. in $\bar{\Omega}$. From (4.9) the pointwise convergence

$$w_h(x, y) \to w(x, y) \quad \text{for all} \quad (x, y) \in \bar{\Omega}$$

follows. Let us suppose that there exists $(x^*, y^*) \in \bar{\Omega}$ such that

$$w(x^*, y^*) < \psi(x^*, y^*)\,. \tag{4.10}$$

As w, ψ are continuous in $\bar{\Omega}$, (4.10) holds in a neighbourhood $\mathcal{U}((x^*, y^*), \varepsilon) \cap \bar{\Omega}$, where $\mathcal{U}((x^*, y^*), \varepsilon)$ is the ball with the center at (x^*, y^*) and the radius $\varepsilon > 0$. Further, $\operatorname{diam}(R_i) \to 0+$ if $h \to 0+$, therefore there exists $a_{i_0}^{h_0} \in \mathring{\mathfrak{N}}_{h_0}$ such that $a_{i_0}^{h_0} \in \mathcal{U}((x^*, y^*), \varepsilon) \cap \bar{\Omega}$. The assumption (4.2) implies that $a_{i_0}^{h} \in \mathring{\mathfrak{N}}_h$ for every $h \leqq h_0$. The definition of K_h yields

$$w_h(a_{i_0}^{h_0}) \geqq \psi(a_{i_0}^{h_0}) \quad \text{for all} \quad h \leqq h_0\,.$$

Hence necessarily

$$w(a_{i_0}^{h_0}) = \lim_{h \to 0+} w_h(a_{i_0}^{h_0}) \geqq \psi(a_{i_0}^{h_0})\,,$$

which contradicts the above considerations.

A direct consequence of Lemmas 4.1 and 4.2 is

Theorem 4.2: *Let all assumptions of Theorem 3.1 be satisfied. If (4.2), (4.3) hold, then*

$$z_h^0 \to z_{,0} \qquad h \to 0+ \quad \text{in } W\,. \tag{4.11}$$

Proof: (4.11) is a consequence of the abstract convergence theorem (see [3]) and the previous two lemmas.

Finally, we want to say a few words on the numerical solution of $(\mathcal{P}_h)$ itself. As the matrices $[D]$, $[K_i]$, $[K_j]$ are symmetric and the bilinear form a is W-elliptic, $(\mathcal{P})$ and $(\mathcal{P}_h)$ can be formulated equivalently as the minimization problem for the quadratic functional

$$\mathcal{J}(z) = 1/2\, a(z, z) - l(z)$$

over K and K_h, respectively. Thus, classical methods of quadratic programming can be employed for the computation of $(\mathcal{P}_h)$, see [3], [12].

References

1 BERNARD, M., Thése, Université Paris 1978.
2 BOCK, I.; LOVÍŠEK, J., On the existence of a weak solution of the boundary value problem for the equilibrium of a shallow shell reinforced with stiffening ribs, Apl. mat. **2** (1978), 132–149.
3 CÉA, J., Optimisation, théorie et algorithmes, Paris, Dunod 1971.
4 CIARLET, P. G., The finite element method for elliptic problems, North-Holland 1978.
5 DUVAUT, G.; LIONS, J. L., Inequalities in mechanics and physics, Springer-Verlag, Berlin-Heidelberg-New York 1976.
6 HASLINGER, J., On numerical solution of a variational inequality of the 4-th order by finite element method, Apl. Mat. **5** (1978), 334–345.
7 HASLINGER, J., Sur la solution d'un problème de la plaque, Apl. Mat. **19** (1974), 316–326.
8 HLAVÁČEK, I.; NEČAS, J., On inequalities of Korn's type in boundary value problems for elliptic systems of partial differential equations, Arch. Rat. Mech. Analysis **36** (1970), 305–311.
9 NAZAROV, A. G., Foundations of the Theory and Method of Computing Shallow Shells (in Russian), Moscow 1966.
10 NEČAS, J., Les méthodes directes en théorie des équations elliptiques, Academia, Prague 1967.
11 SEIDE, P., Small Elastic Deformations of Thin Shells, Nordhoff International Publishing, London 1975.
12 GLOWINSKI, R.; LIONS, J. L.; TRÉMOLIERES, R., Analyse numérique des inéquations variationnelles, Paris, Dunod 1976.

Eingereicht: 10. 9. 1980

Anschriften: Dr. JAROSLAV HASLINGER, Matematicko-fyzikalni fakulta, Univerzita Karlova, Malostranské nám. 25 Praha 1, 11800 Československo; Dr. JAN LOVÍŠEK, SvF SVŠT, Bratislava, 88420 Radlinskeho 11, Československo

ZAMM 62, 35–41 (1982)

D. IEŞAN

Some Theorems in a Grade Consistent Micropolar Theory of Elasticity

In der vorliegenden Arbeit wird eine verallgemeinerte mikropolare Elastizitätstheorie behandelt. Es werden einige allgemeine Theoreme, insbesondere Existenz- und Eindeutigkeits-, Reziprozitäts- und Variationstheoreme hergeleitet.

A grade-consistent micropolar theory of elasticity is considered. Some general theorems (existence and uniqueness theorems, reciprocal theorems, variational theorems) are established.

В настоящей работе рассматривается обобщённой микрополярной теории упругости. Приводятся некоторые теоремы, а именно: теорема об однозначности и существования, теорема о взаимности и вариационная теорема.

1. Introduction

The theory of micropolar elasticity was intensively studied during recent years (see e.g. [1]–[6]). Recently, BRULIN and HJALMARS [7] have derived a micropolar theory where the second-order displacement gradient is added to the classic set of independent constitutive variables. This theory was named the grade-consistent micropolar theory. In [7], an analysis of constitutive equations of the non-linear thermoelastic media is presented. The field equations of the linear theory of elasticity were considered in [7], [8].

The present paper is concerned with the linear theory of elastic solids. The purpose of this paper is to prove basic theorems concerning the existence and uniqueness of solution, the reciprocity relations and the variational characterization of the solution. In order to study the boundary conditions, in the first part of the paper we present a derivation of the linear theory based on the method given in [9]. In the second part of the paper we establish general theorems.

2. Basic Equations

We refer the motion of the continuum to a fixed system of rectangular Cartesian axes $0x_i$ $(i = 1, 2, 3)$. We shall employ the usual summation and differentiation conventions: Latin subscripts (unless otherwise specified) are understood to range over the integers (1, 2, 3), summation over repeated subscripts is implied and subscripts preceded by a comma denote partial differentiation with respect to the corresponding Cartesian coordinate.

In what follows we consider the linear theory of mechanics of continuous media, each particle having six degrees of freedom. We denote by u_i the components of the displacement vector and by φ_i the components of the microrotation vector.

We consider an arbitrary material volume $\mathcal{V}$ in the continuum bounded by a surface $\mathcal{A}$ at time t, and we suppose that V is the corresponding volume in the initial undeformed state of the continuum, bounded by a surface A. Let the outward unit normal at A be n_i, referred to our fixed rectangular frame of reference. We postulate an energy balance for an arbitrary material volume, in the form

$$\int_V (\varrho v_i \dot{v}_i + I_{ij} \nu_i \dot{\nu}_j) \,\mathrm{d}V + \int_V \varrho \dot{U} \,\mathrm{d}V = \int_V (F_i v_i + M_i \nu_i + F_{ji} v_{i,j}) \,\mathrm{d}V + \int_A (t_i v_i + m_i \nu_i + \mu_{ji} v_{i,j}) \,\mathrm{d}A \,, \quad (2.1)$$

where $v_i = \dot{u}_i$, $\nu_i = \dot{\varphi}_i$, ϱ is the density in the reference configuration, I_{ij} $(= I_{ji})$ are coefficients of inertia, U is the internal energy per unit mass, F_i is the body force per unit volume in the reference configuration, M_i is the body couple per unit volume in the reference configuration, F_{ij} is the dipolar body force, t_i is a part of the stress vector associated with the surface $\mathcal{A}$ but measured per unit area of the surface A, m_i is the couple stress, μ_{ij} is the dipolar surface force, and a superposed dot denotes the material derivative with respect to the time. We suppose that the body has arrived at a given state at a time t through some prescribed motion. Following [9], we consider a second motion which differs from the given motion only by a constant superposed rigid body translational velocity, the body occupying the same position at time t, and we assume that $\dot{U}$, F_i, M_i, F_{ij}, t_i, m_i, μ_{ij} are unaltered by such superposed rigid body velocity. If we use equation (2.1) with v_i replaced by $v_i + a_i$, where a_i is an arbitrary constant, we obtain

$$\left[\int_V \varrho \dot{v}_i \,\mathrm{d}V - \int_V F_i \,\mathrm{d}V - \int_A t_i \,\mathrm{d}A\right] a_i = 0 \,,$$

for all arbitrary constants a_i. The quantities in the square brackets are independent of a_i so that we obtain

$$\int_V \varrho \dot{v}_i \,\mathrm{d}V = \int_V F_i \,\mathrm{d}V + \int_A t_i \,\mathrm{d}A \,. \quad (2.2)$$

We now take a volume $\mathcal{V}$ which is such that in V it was a tetrahedral element bounded by a plane with arbitrary unit normal n_i, and by planes through the point x_i, parallel to the coordinate planes. Using the well-known method, we obtain

$$t_i = t_{ji} n_j , \tag{2.3}$$

and

$$t_{ji,j} + F_i = \varrho \ddot{u}_i . \tag{2.4}$$

Taking into account (2.3) and (2.4), the equation (2.1) reduces to

$$\int_V I_{ij} \nu_i \dot{\nu}_j \, dV + \int_V \varrho \dot{U} \, dV = \int_V (M_i \nu_i + F_{ji} v_{i,j}) \, dV + \int_V t_{ji} v_{i,j} \, dV + \int_A (m_i \nu_i + \mu_{ji} v_{i,j}) \, dA . \tag{2.5}$$

With an argument similar to that used in obtaining (2.3), we have

$$(m_i - m_{ji} n_j) \nu_i + (\mu_{ji} - \mu_{rji} n_r) v_{i,j} = 0 , \tag{2.6}$$

where m_{ji} is the couple stress tensor and μ_{rji} is the hyperstress tensor. With the help of (2.6), equation (2.5) reduces to

$$(m_{ji,j} + M_i - I_{ij} \dot{\nu}_j) \nu_i + (\mu_{kji,k} + t_{ji} + F_{ji}) v_{i,j} + m_{ji} \nu_{i,j} + \mu_{kji} v_{i,jk} - \varrho \dot{U} = 0 . \tag{2.7}$$

Let us now consider a motion of the body which differs from the given motion only by a superposed uniform rigid body angular velocity, the body occupying the same position at time t, and let us assume that $\dot{U}$, M_i, F_{ij}, t_{ij}, m_{ij}, μ_{ijk} are unaltered by such motion. In this case ν_i are replaced by $\nu_i + b_i$, where b_i are arbitrary constants, and v_i are replaced by $v_i + \varepsilon_{ijk} b_j x_k$, where ε_{ijk} is the alternating symbol. Equation (2.7) holds when $v_{i,j}$ is replaced by $v_{i,j} + \varepsilon_{jik} b_k$ and ν_i by $\nu_i + b_i$. It follows

$$(m_{ji,j} + M_i - I_{ij} \dot{\nu}_j) (\nu_i + b_i) + (\mu_{kji,k} + t_{ji} + F_{ji}) \cdot (v_{i,j} + \varepsilon_{jir} b_r) + m_{ji} \nu_{i,j} + \mu_{kji} v_{i,jk} - \varrho \dot{U} = 0 .$$

Using (2.7), we obtain the equations

$$m_{ji,j} + \varepsilon_{irs} \tau_{rs} + M_i = I_{ij} \ddot{\varphi}_j , \tag{2.8}$$

where

$$\tau_{rs} = \mu_{krs,k} + t_{rs} + F_{rs} . \tag{2.9}$$

With the help of (2.8), the equation (2.7) reduces to

$$\varrho \dot{U} = \tau_{ij} \dot{e}_{ij} + m_{ij} \dot{\varkappa}_{ij} + \mu_{ijk} \dot{\varkappa}_{ijk} , \tag{2.10}$$

where we have used the notations

$$e_{ij} = u_{j,i} + \varepsilon_{jik} \varphi_k , \qquad \varkappa_{ij} = \varphi_{j,i} , \qquad \varkappa_{ijk} = u_{k,ij} . \tag{2.11}$$

We restrict attention only to the linear theory of elastic materials where

$$U = U(e_{rs}, \varkappa_{rs}, \varkappa_{rpq}) , \qquad t_{ij} = t_{ij}(e_{rs}, \varkappa_{rs}, \varkappa_{rpq}) , \tag{2.12}$$

$$m_{ij} = m_{ij}(e_{rs}, \varkappa_{rs}, \varkappa_{rpq}) , \qquad \mu_{ijk} = \mu_{ijk}(e_{rs}, \varkappa_{rs}, \varkappa_{rpq}) ,$$

the functionals in (2.12) being consistent with the assumption of the linear theory. It is easy to see that e_{ij}, $\varkappa_{ij}$, $\varkappa_{ijk}$ are invariant under superposed rigid-body motions.

For a given deformation ν_i and $v_{i,j}$ in (2.6) may be chosen arbitrarily and independently of each other so that

$$m_i = m_{ji} n_j , \qquad \mu_{ij} = \mu_{rij} n_r . \tag{2.13}$$

From (2.11) and (2.12) it follows

$$\tau_{ij} = \varrho \frac{\partial U}{\partial e_{ij}} , \qquad m_{ij} = \varrho \frac{\partial U}{\partial \varkappa_{ij}} , \qquad \mu_{(ij)k} = \varrho \frac{\partial U}{\partial \varkappa_{ijk}} , \tag{2.14}$$

where $\mu_{(ij)k}$ is the part of μ_{ijk}, symmetric with respect to i, j. The skew symmetric part $\mu_{[ij]k}$ makes no contribution to the rate of work over any closed surface in the body, or over the complete boundary of the body. Thus, with the help of (2.3), (2.9), (2.13), the surface integral in (2.1) becomes

$$\int_A (t_i v_i + m_i \nu_i + \mu_{ji} v_{i,j}) \, dA = \int_A (t_{ji} v_i + m_{ji} \nu_i + \mu_{jrs} v_{s,r}) \, n_j \, dA$$

$$= \int_A [(\tau_{ji} - F_{ji} - \mu_{sji,s}) v_i + m_{ji} \nu_i + \mu_{jrs} v_{s,r}] \, n_j \, dA$$

$$= \int_A [(\tau_{ji} - F_{ji} - \mu_{(sj)i,s}) v_i + m_{ji} \nu_i + \mu_{(jr)s} v_{s,r}] \, n_j \, dA + \int_A (\mu_{[js]r,s} v_r + \mu_{[js]r} v_{r,s}) \, n_j \, da$$

$$= \int_A [(\tau_{ji} - F_{ji} - \mu_{(sj)i,s}) v_i + m_{ji} \nu_i + \mu_{(jr)s} v_{s,r}] \, n_j \, dA + \int_V (\mu_{[js]r} v_r)_{,js} \, dV$$

$$= \int_A [(\tau_{ji} - F_{ji} - \mu_{(sj)i,s}) v_i + m_{ji} \nu_i + \mu_{(jr)s} v_{s,r}] \, n_j \, dA .$$

It is easy to see that $\mu_{[ij]k}$ makes no contribution to the equations of motions. In what follows we assume that $\mu_{[ij]k} = 0$ (to this conclusion we can arrive following the method used in [10], [11]).

In the linear theory, we have

$$W \equiv \varrho U = \tfrac{1}{2} A_{ijrs} e_{ij} e_{rs} + B_{ijrs} e_{ij} \varkappa_{rs} + \tfrac{1}{2} C_{ijrs} \varkappa_{ij} \varkappa_{rs} + D_{ijpqr} e_{ij} \varkappa_{pqr} + E_{ijpqr} \varkappa_{ij} \varkappa_{pqr} + \tfrac{1}{2} F_{ijkpqr} \varkappa_{ijk} \varkappa_{pqr}, \tag{2.15}$$

where

$$A_{ijrs} = A_{rsij}, \quad C_{ijrs} = C_{rsij}, \quad D_{ijpqr} = D_{ijqpr}, \quad E_{ijpqr} = E_{ijqpr}, \quad F_{ijkpqr} = F_{jikpqr} = F_{pqrijk}. \tag{2.16}$$

From (2.14) and (2.15) we obtain

$$\begin{aligned} \tau_{ij} &= A_{ijrs} e_{rs} + B_{ijrs} \varkappa_{rs} + D_{ijpqr} \varkappa_{pqr}, \\ m_{ij} &= B_{rsij} e_{rs} + C_{ijrs} \varkappa_{rs} + E_{ijpqr} \varkappa_{pqr}, \\ \mu_{ijk} &= D_{rsijk} e_{rs} + E_{rsijk} \varkappa_{rs} + F_{ijkpqr} \varkappa_{pqr}. \end{aligned} \tag{2.17}$$

Let us consider the question of boundary conditions. We assume that the body occupies a bounded region B with Lipschitz boundary ∂B which consists of a finite number of smooth surfaces. Let γ_α be intersections of two adjoined smooth surfaces and $\Gamma = \bigcup\limits_{\alpha} \gamma_\alpha$. In the last integral of (2.1), $v_{i,j}$ is not independent of v_i on A: only the normal component $Dv_i = v_{i,s} n_s$ is independent. We separate latter

$$\mu_{ji} v_{i,j} = \mu_{pji} n_p (n_j D v_i + \varepsilon_{mrl} \varepsilon_{sjl} n_s n_m v_{i,r}) = \mu_{sji} n_s n_j D v_i + \varepsilon_{mrl} n_m (\varepsilon_{sjl} n_s n_p \mu_{pji} v_i)_{,r} - \varepsilon_{mrl} n_m v_i (\varepsilon_{sjl} n_s n_p \mu_{pji})_{,r}.$$

By Stokes's theorem we have

$$\int_{\partial B} \mu_{ji} v_{i,j} \, dA = \int_{\partial B} \mu_{sji} n_s n_j D v_i \, dA + \int_{\partial B} (\mu_{pji} n_p n_r - \mu_{pri} n_p n_j)_{,r} v_i n_j \, dA + \int_{\Gamma} Q_i v_i \, ds, \tag{2.18}$$

where

$$Q_i = \langle \mu_{pji} n_p \eta_j \rangle, \qquad \eta_j = \varepsilon_{jrs} s_r n_s; \tag{2.19}$$

$\langle\,\rangle$ denotes the difference of limits from both sides of γ_α and s_i represents the unit tangent vector to γ_α. With the help of (2.18) the total rate of work over the boundary of the body becomes

$$\int_{\partial B} (t_i v_i + m_i \nu_i + \mu_{ji} v_{i,j}) \, dA = \int_{\partial B} (P_i v_i + m_i \nu_i + R_i D v_i) \, dA + \int_{\Gamma} Q_i v_i \, ds, \tag{2.20}$$

where

$$\begin{aligned} P_i &= [t_{ji} + (\mu_{pji} n_p n_r - \mu_{pri} n_p n_j)_{,r}] \, n_j = \\ &= [\tau_{ji} - F_{ji} - \mu_{sji,s} + (\mu_{pji} n_p n_r - \mu_{pri} n_p n_j)_{,r}] \, n_j, \\ R_i &= \mu_{rsi} n_r n_s. \end{aligned} \tag{2.21}$$

Let us consider the subsets Σ_s $(s = 1, 2, \dots, 6)$ of ∂B so that $\Sigma_1 \cup \Sigma_2 \cup \Gamma = \Sigma_3 \cup \Sigma_4 \cup \Gamma = \Sigma_5 \cup \Sigma_6 \cup \Gamma = \partial B$ $\Sigma_1 \cap \Sigma_2 = \Sigma_3 \cap \Sigma_4 = \Sigma_5 \cap \Sigma_6 = 0$. We consider the following boundary conditions

$$\begin{aligned} &u_i = \tilde{u}_i \ \text{ on } \ \Sigma_1 \times [0, t_0), \qquad P_i = \tilde{P}_i \ \text{ on } \ \Sigma_2 \times [0, t_0), \\ &\varphi_i = \tilde{\varphi}_i \ \text{ on } \ \Sigma_3 \times [0, t_0), \qquad m_i = \tilde{m}_i \ \text{ on } \ \Sigma_4 \times [0, t_0), \\ &D u_i = g_i \ \text{ on } \ \Sigma_5 \times [0, t_0), \qquad R_i = \tilde{R}_i \ \text{ on } \ \Sigma_6 \times [0, t_0), \\ &Q_i = \tilde{Q}_i \ \text{ on } \ \Gamma \times [0, t_0), \end{aligned} \tag{2.22}$$

where $\tilde{u}_i$, $\tilde{P}_i$, $\tilde{\varphi}_i$, $\tilde{m}_i$, g_i, $\tilde{R}_i$, $\tilde{Q}_i$ are prescribed functions and t_0 is some time instant that may be infinite. With the help of (2.9), the equations (2.4) become

$$\tau_{ji,j} - \mu_{sji,sj} - F_{ji,j} + F_i = \varrho \ddot{u}_i. \tag{2.23}$$

The basic equations of the linear theory are: the equations of motion (2.8), (2.23), the constitutive equations (2.17), and the geometrical equations (2.11).

To the system of field equations we adjoin the boundary conditions (2.22) and the initial conditions

$$\begin{aligned} &u_i(x, 0) = a_i(x), \qquad \dot{u}_i(x, 0) = b_i(x), \\ &\varphi_i(x, 0) = c_i(x), \qquad \dot{\varphi}_i(x, 0) = d_i(x), \qquad x \in \bar{B}, \end{aligned} \tag{2.24}$$

where a_i, b_i, c_i, d_i are prescribed functions.

3. The Linear Theory of Elastostatics

In the case of equilibrium, the equations (2.8) and (2.23) become

$$\tau_{ji,j} - \mu_{sji,sj} + F_i - F_{ji,j} = 0, \qquad m_{ji,j} + \varepsilon_{irs} \tau_{rs} + M_i = 0. \tag{3.1}$$

Let us consider the body subjected to two different systems of loadings $L^{(\alpha)} = \{F_i^{(\alpha)}, M_i^{(\alpha)}, F_{ij}^{(\alpha)}, \tilde{u}_i^{(\alpha)}, \tilde{P}_i^{(\alpha)}, \tilde{\varphi}_i^{(\alpha)}, \tilde{m}_i^{(\alpha)}, g_i^{(\alpha)}, \tilde{R}_i^{(\alpha)}, \tilde{Q}_i^{(\alpha)}\}$, $(\alpha = 1, 2)$, and the corresponding states $C^{(\alpha)} = \{u_i^{(\alpha)}, \varphi_i^{(\alpha)}, e_{ij}^{(\alpha)}, \varkappa_{ij}^{(\alpha)}, \varkappa_{ijk}^{(\alpha)}, \tau_{ij}^{(\alpha)}, m_{ij}^{(\alpha)}, \mu_{ijk}^{(\alpha)}\}$. We denote by $P_i^{(\alpha)}$, $m_i^{(\alpha)}$, $R_i^{(\alpha)}$, $Q_i^{(\alpha)}$ the values of P_i, m_i, R_i, Q_i corresponding to the states $C^{(\alpha)}$.

Reciprocal theorem. *If an elastic solid is subjected to two systems of loadings* $L^{(\alpha)}$ $(\alpha = 1, 2)$, *then between the corresponding states* $C^{(\alpha)}$ *there is the following relation*

$$\int_B [(F_i^{(1)} - F_{ji,j}^{(1)})\, u_i^{(2)} + M_i^{(1)}\varphi_i^{(2)}]\,\mathrm{d}V + \int_{\partial B} [(P_i^{(1)} + F_{ji}^{(1)} n_j)\, u_i^{(2)} + m_i^{(1)}\varphi_i^{(2)} + R_i^{(1)} D u_i^{(2)}]\,\mathrm{d}A + \int_\Gamma Q_i^{(1)} u_i^{(2)}\,\mathrm{d}s$$
$$= \int_B [(F_i^{(2)} - F_{ji,j}^{(2)})\, u_i^{(1)} + M_i^{(2)}\varphi_i^{(1)}]\,\mathrm{d}V + \int_{\partial B} [(P_i^{(2)} + F_{ji}^{(2)} n_j)\, u_i^{(1)} + m_i^{(2)}\varphi_i^{(1)} + R_i^{(2)} D u_i^{(1)}]\,\mathrm{d}A +$$
$$+ \int_\Gamma Q_i^{(2)} u_i^{(1)}\,\mathrm{d}A\,. \tag{3.2}$$

Proof. On the basis of the relations (2.16), we conclude from the constitutive equations that

$$[\tau_{ij}^{(1)} - B_{ijrs}\varkappa_{rs}^{(1)} - D_{ijpqr}\varkappa_{pqr}^{(1)}]\, e_{ij}^{(2)} = [\tau_{ij}^{(2)} - B_{ijrs}\varkappa_{rs}^{(2)} - D_{ijpqr}\varkappa_{pqr}^{(2)}]\, e_{ij}^{(1)}\,,$$
$$[m_{ij}^{(1)} - B_{rsij}e_{rs}^{(1)} - E_{ijpqr}\varkappa_{pqr}^{(1)}]\, \varkappa_{ij}^{(2)} = [m_{ij}^{(2)} - B_{rsij}e_{rs}^{(2)} - E_{ijpqr}\varkappa_{pqr}^{(2)}]\, \varkappa_{ij}^{(1)}\,,$$
$$[\mu_{ijk}^{(1)} - D_{rsijk}e_{rs}^{(1)} - E_{rsijk}\varkappa_{rs}^{(1)}]\, \varkappa_{ijk}^{(2)} = [\mu_{ijk}^{(2)} - D_{rsijk}e_{rs}^{(2)} - E_{rsijk}\varkappa_{rs}^{(2)}]\, \varkappa_{ijk}^{(1)}\,.$$

Adding up these relations we obtain

$$\tau_{ij}^{(1)}e_{ij}^{(2)} + m_{ij}^{(1)}\varkappa_{ij}^{(2)} + \mu_{ijk}^{(1)}\varkappa_{ijk}^{(2)} = \tau_{ij}^{(2)}e_{ij}^{(1)} + m_{ij}^{(2)}\varkappa_{ij}^{(1)} + \mu_{ijk}^{(2)}\varkappa_{ijk}^{(1)}\,. \tag{3.3}$$

If we introduce the notations

$$K_{\alpha\beta} = \int_B (\tau_{ij}^{(\alpha)}e_{ij}^{(\beta)} + m_{ij}^{(\alpha)}\varkappa_{ij}^{(\beta)} + \mu_{ijk}^{(\alpha)}\varkappa_{ijk}^{(\beta)})\,\mathrm{d}V\,, \tag{3.4}$$

then, from (3.3), we have

$$K_{12} = K_{21}\,. \tag{3.5}$$

By using the relations (2.11), (2.18), (3.1) and the divergence theorem, we get

$$K_{\alpha\beta} = \int_{\partial B} [(\tau_{ji}^{(\alpha)} - \mu_{sji,s}^{(\alpha)})\, n_j u_i^{(\beta)} + m_i^{(\alpha)}\varphi_i^{(\beta)} + \mu_{sji}^{(\alpha)} u_{i,j}^{(\beta)} n_s]\,\mathrm{d}A + \int_B [(F_i^{(\alpha)} - F_{ji,i}^{(\alpha)})\, u_i^{(\beta)} + M_i^{(\alpha)}\varphi_i^{(\beta)}]\,\mathrm{d}V \tag{3.6}$$
$$= \int_{\partial B} [(P_i^{(\alpha)} + F_{ji}^{(\alpha)} n_j)\, u_i^{(\beta)} + m_i^{(\alpha)}\varphi_i^{(\beta)} + R_i^{(\alpha)} D u_i^{(\beta)}]\,\mathrm{d}A + \int_\Gamma Q_i^{(\alpha)} u_i^{(\beta)}\,\mathrm{d}s + \int_B [(F_i^{(\alpha)} - F_{ji,j}^{(\alpha)})\, u_i^{(\beta)}$$
$$+ M_i^{(\alpha)}\varphi_i^{(\beta)}]\,\mathrm{d}V\,.$$

From (3.5) and (3.6) we obtain (3.2).

As in classical theory of elasticity we can obtain various applications of the reciprocal theorem.

Let us note that if we take $L^{(1)} = \{F_i, M_i F_{ij}, \tilde{u}_i, \tilde{P}_i, \tilde{\varphi}_i, \tilde{m}_i, g_i, \tilde{R}_i, \tilde{Q}_i\}$ and $C^{(1)} = \{u_i, \varphi_i, e_{ij}, \varkappa_{ij}, \varkappa_{ijk}, \tau_{ji}, m_{ij}, \mu_{ijk}\}$, then, from (2.15), (2.17), (3.4) we obtain

$$K_{11} = 2 \int_B W\,\mathrm{d}V\,,$$

and from (3.6) we get

$$K_{11} = \int_B [(F_i - F_{ji,j})\, u_i + M_i\varphi_i]\,\mathrm{d}V + \int_{\partial B} [(P_i + F_{ji}n_j)\, u_i + m_i\varphi_i + R_i D u_i]\,\mathrm{d}A + \int_\Gamma Q_i u_i\,\mathrm{d}s\,.$$

Thus

$$2 \int_B W\,\mathrm{d}V = \int_B [(F_i - F_{ji,j})\, u_i + M_i\varphi_i]\,\mathrm{d}V + \int_{\partial B} [(P_i + F_{ji}n_j)\, u_i + m_i\varphi_i + R_i D u_i]\,\mathrm{d}A + \int_\Gamma Q_i u_i\,\mathrm{d}s\,. \tag{3.7}$$

Uniqueness theorem. *Suppose that the form W is positive definite. Then any two solutions of the mixed problem are equal modulo a rigid displacement. Moreover, if* Σ_1 *is non-empty, then the mixed problem has at most one solution.*

Proof. Let $\{u_i, \varphi_i, e_{ij}, \varkappa_{ij}, \varkappa_{ijk}, \tau_{ij}, m_{ij}, \mu_{ijk}\}$ and $\{\bar{u}_i, \bar{\varphi}_i, \bar{e}_{ij}, \bar{\varkappa}_{ij}, \bar{\varkappa}_{ijk}; \bar{\tau}_{ij}, \bar{m}_{ij}, \bar{\mu}_{ijk}\}$ be solutions of the mixed problem, and let $u_i^0 = u_i - \bar{u}_i$, $\varphi_i^0 = \varphi_i - \bar{\varphi}_i, \ldots, \mu_{ijk}^0 = \mu_{ijk} - \bar{\mu}_{ijk}$. According to the linearity of the problem, $\{u_i^0, \varphi_i^0, e_{ij}^0, \varkappa_{ij}^0, \varkappa_{ijk}^0, \tau_{ij}^0, m_{ij}^0, \mu_{ijk}^0\}$ is a solution corresponding to $F_i = 0$, $F_{ij} = 0$, $M_i = 0$. Moreover we have

$$u_i^0 = 0 \text{ on } \Sigma_1\,, \quad P_i^0 = 0 \text{ on } \Sigma_2\,, \quad \varphi_i^0 = 0 \text{ on } \Sigma_3\,, \quad m_i^0 = 0 \text{ on } \Sigma_4\,,$$
$$Du_i^0 = 0 \text{ on } \Sigma_5\,, \quad R_i^0 = 0 \text{ on } \Sigma_6\,, \quad Q_i^0 = 0 \text{ on } \Gamma\,,$$

thus

$$P_i^0 u_i^0 + m_i^0\varphi_i^0 + R_i^0 D u_i^0 = 0 \quad \text{on } \partial B\,,$$

and we conclude from (3.7) that

$$\int_B W_0\,\mathrm{d}V = 0\,, \tag{3.8}$$

where W^0 is the strain energy corresponding to strain fields e_{ij}^0, $\varkappa_{ij}^0$, $\varkappa_{ijk}^0$. From (3.8) we obtain $e_{ij}^0 = 0$, $\varkappa_{ij}^0 = 0$, $\varkappa_{ijk}^0 = 0$ and therefore

$$u_i = \alpha_i + \varepsilon_{ijk}\beta_j x_k\,, \quad \varphi_i^0 = \beta_i\,, \tag{3.9}$$

where α_i and β_i are arbitrary constants. If Σ_1 is non-empty we obtain $u_i^0 = 0$, $\varphi_i^0 = 0$ in B.

The fundamental system of field equations for the time-independent behavior of a linear elastic body consists of the geometrical equations (2.11), the constitutive equations (2.17) and the equations of equilibrium (3.1). It is easy to derive a characterization of the equilibrium equations in terms of displacements and microrotations. Thus, we obtain the equations

$$L_i u = F_i - F_{ji,j}\,, \qquad L_{3+i} u = M_i\,,$$

where $\boldsymbol{u} = (u_1, u_2, u_3, \varphi_1, \varphi_2, \varphi_3) \equiv (u_i, \varphi_i)$, and the expressions of the operators L_s can be found easily. If we denote

$$\boldsymbol{L}\boldsymbol{u} = (L_1\boldsymbol{u}, L_2\boldsymbol{u}, \ldots, L_6\boldsymbol{u})\,, \qquad \boldsymbol{f} = (F_i - F_{ji,j}, M_i)\,,$$

the above equations can be written in the form

$$\boldsymbol{L}\boldsymbol{u} = \boldsymbol{f}\,. \tag{3.10}$$

Let us consider the vectors $\boldsymbol{a} = (a_1, a_2, \ldots, a_6)$, $\boldsymbol{b} = (b_1, b_2, \ldots, b_6)$. We will denote by $\boldsymbol{ab}$ the scalar product $\boldsymbol{ab} = \sum_1^6 a_s b_s$. If we introduce the notations

$$\begin{aligned} &\boldsymbol{u} = (u_i^{(1)}, \varphi_i^{(1)}), \quad \boldsymbol{v} = (u_i^{(2)}, \varphi_i^{(2)})\,, \quad K(\boldsymbol{u}, \boldsymbol{v}) = K_{12}\,, \quad p_i(\boldsymbol{u}) = P_i^{(1)} + F_{ji}^{(1)} n_j\,, \\ &m_i(\boldsymbol{u}) = m_i^{(1)}\,, \quad r_i(\boldsymbol{u}) = R_i^{(1)}\,, \quad q_i(u) = Q_i^{(1)}\,, \quad v_i = u_i^{(2)}\,, \quad \psi_i = \varphi_i^{(2)}\,, \end{aligned} \tag{3.11}$$

then, from (3.4) and (3.6) we obtain

$$\int_B \boldsymbol{v}\boldsymbol{L}\boldsymbol{u}\,\mathrm{d}V = K(\boldsymbol{u}, \boldsymbol{v}) - \int_{\partial B} [v_i p_i(\boldsymbol{u}) + \psi_i m_i(\boldsymbol{u}) + r_i(\boldsymbol{u})\,\mathrm{D}v_i]\,\mathrm{d}A + \int_\Gamma v_i q_i(\boldsymbol{u})\,\mathrm{d}s\,.$$

We note that the relation (3.5) implies $K(\boldsymbol{u}, \boldsymbol{v}) = K(\boldsymbol{v}, \boldsymbol{u})$. In the case of homogeneous boundary conditions the operator L is symmetric.

Following [12]—[15] we can obtain

Theorem of existence. *Suppose that the internal energy is a positive definite form.*

i) *Let Σ_1, Σ_3 and Σ_5 be empty. The necessary and sufficient conditions for the existence of a solution are*

$$\int_B F_i\,\mathrm{d}V + \int_{\partial B} \tilde{P}_i\,\mathrm{d}A + \int_\Gamma \tilde{Q}_i\,\mathrm{d}s = 0\,, \tag{3.12}$$

$$\int_B [\varepsilon_{ijk} x_j (F_k - F_{sk,s}) + M_i]\,\mathrm{d}V + \int_{\partial B} \{\varepsilon_{ijk}[x_j(\tilde{P}_k + F_{sk} n_s) + n_j \tilde{R}_k] + \tilde{m}_i\}\,\mathrm{d}A + \int_\Gamma \varepsilon_{ijk} x_j \tilde{Q}_k\,\mathrm{d}s = 0\,.$$

ii) *Let Σ_1 contain a non-empty set open in ∂B. Then there exists one and only one solution.*

We say that $S = \{u_i, \varphi_i, e_{ij}, \varkappa_{ij}, \varkappa_{ijk}, \tau_{ij}, m_{ij}, \mu_{ijk}\}$ is a kinematically admissible state if

i) $u_i \in C^2(B)$, $\varphi_i \in C^2(B)$, $u_i \in C^1(\bar{B})$, $\varphi_i \in C^1(\bar{B})$;

ii) the functions $u_i, \varphi_i, e_{ij}, \varkappa_{ij}, \varkappa_{ijk}, \tau_{ij}, m_{ij}, \mu_{ijk}$ satisfy the equations (2.11), (2.17) and the boundary conditions imposed on Σ_1, Σ_3 and Σ_5.

Principle of minimum potential energy. *Let K denote the set of all kinematically admissible states, and let Λ be the functional on K defined by*

$$\begin{aligned} \Lambda(S) = &\int_B [W - (F_i - F_{ji,j})\,u_i - M_i\varphi_i]\,\mathrm{d}V - \\ &- \int_{\Sigma_2} (\tilde{P}_i + F_{ji} n_j)\,u_i\,\mathrm{d}A - \int_{\Sigma_4} \tilde{m}_i\varphi_i\,\mathrm{d}A - \int_{\Sigma_6} \tilde{R}_i D u_i\,\mathrm{d}A - \int_\Gamma \tilde{Q}_i u_i\,\mathrm{d}s\,, \end{aligned} \tag{3.13}$$

for every $S = \{u_i, \varphi_i, e_{ij}, \varkappa_{ij}, \varkappa_{ijk}, \tau_{ij}, m_{ij}, \mu_{ijk}\} \in K$, where W is given by (2.15). Let S^* be the solution of the mixed problem. Then

$$\Lambda(S^*) \leqq \Lambda(S)\,, \tag{3.14}$$

for every $S \in K$, and equality holds only if $S^* = S$ modulo a rigid displacement.

The proof of this theorem can be made by the procedure used in the classical theory of elasticity [16].

4. The Dynamic Theory of Elasticity

Uniqueness theorem. *Let W be positive semi-definite, I_{ij} be a positive definite tensor and ϱ be strictly positive. Then the mixed problem has at most one solution.*

Proof. On the basis of the equation (2.8), (2.11), (2.23), the relation (2.10) can be expressed as

$$\begin{aligned} \dot{W} &= \tau_{ji}(v_{i,j} + \varepsilon_{kij}\nu_k) + m_{ji}\nu_{i,j} + \mu_{sji} v_{i,js} \\ &= (\tau_{ji} v_i)_{,j} + (m_{ji}\nu_i)_{,j} + (\mu_{sji} v_{i,j})_{,s} - (\mu_{sji,s} v_i)_{,j} - (\tau_{ji,j} - \mu_{sji,sj})\,v_i - (m_{ji,j} + \varepsilon_{isr}\tau_{rs})\,\nu_i \\ &= [(\tau_{ji} - \mu_{sji,s})\,v_i]_{,j} + (m_{ji}\nu_i)_{,j} + (\mu_{sji} v_{i,j})_{,s} + (F_i - F_{ji,j})\,v_i + M_i\nu_i - \varrho v_i \dot{v}_i - I_{ij}\dot{\nu}_j \nu_i\,. \end{aligned} \tag{4.1}$$

We call the function E on $[0, t_0)$ defined by

$$E = \frac{1}{2} \int_B (\varrho v_i v_i + I_{ij}\nu_i\nu_j)\,\mathrm{d}V + \int_B W\,\mathrm{d}V\,,$$

the total energy. Using the divergence theorem and (2.20), from (4.1) we obtain

$$\dot{E} = \int_B [(F_i v_i + F_{ji} v_{i,j} + M_i \nu_i] \, dV + \int_{\partial B} (P_i v_i + m_i \nu_i + R_i D v_i) \, dA + \int_\Gamma Q_i v_i \, ds \,. \tag{4.2}$$

Let $S^0 = \{u_i^0, \varphi_i^0, e_{ij}^0, \varkappa_{ij}^0, \varkappa_{ijk}^0, \tau_{ij}^0, m_{ij}^0, \mu_{ijk}^0\}$ be the difference between two solutions. According to the linearity of the problem, S^0 is a solution corresponding to $F_i = 0$, $M_i = 0$, $F_{ij} = 0$ and satisfies

$$u_i^0(x, 0) = 0 \,, \quad \dot{u}_i^0(x, 0) = 0 \,, \quad \varphi_i^0(x, 0) = 0 \,, \quad \dot{\varphi}_i^0(x, 0) = 0 \,, \quad x \in \bar{B} \,, \tag{4.3}$$
$$P_i^0 u_i^0 + m_i^0 \varphi_i^0 + R_i^0 D u_i^0 = 0 \quad \text{on} \quad \partial B, \; Q_i^0 = 0 \quad \text{on} \quad \Gamma \,.$$

By use of (4.2) and $(4.3)_2$ we obtain $E^0(t) = E^0(0)$, $0 \leq t < t_0$. The conditions $(4.3)_1$ imply $E^0(0) = 0$. Thus we conclude

$$E^0(t) = 0 \,, \qquad 0 \leq t < t_0 \,. \tag{4.4}$$

By hypothesis and by (4.4), $\dot{u}_i^0 = 0$, $\dot{\varphi}_i^0 = 0$ on $\bar{B} \times [0, t_0)$. But u_i^0 and φ_i^0 vanish initially; thus $u_i^0 = 0$, $\varphi_i^0 = 0$ on $\bar{B} \times [0, t_0)$.

Alternative formulation of the boundary-initial-value problem. Let u and v be functions defined on $\bar{B} \times [0, \infty)$, continuous on $[0, \infty)$ with respect to the time t for each $x \in \bar{B}$. We denote by $u * v$ the convolution of u and v:

$$[u * v] (x, t) = \int_0^t u(x, t - \tau) \, v(x, \tau) \, d\tau \,.$$

Let us introduce the notations

$$g(t) = t \,, \qquad f_i = g * (F_i - F_{ji,j}) + \varrho(tb_i + a_i) \,, \qquad l_i = g * M_i + I_{ij}(td_j + c_j) \,. \tag{4.5}$$

Proceeding as in [17], [18] one can prove the following theorem:

The functions $u_i, \varphi_i, \tau_{ij}, m_{ij}, \mu_{ijk}$ satisfy the equations (2.8), (2.3) and the initial conditions (2.24) if and only if

$$g * (\tau_{ji,j} - \mu_{sji,sj}) + f_i = \varrho u_i \,, \tag{4.6}$$
$$g * (m_{ji,j} + \varepsilon_{irs} \tau_{rs}) + l_i = I_{ij} \varphi_j \,.$$

This theorem enables us to give an alternate formulation of the boundary-initial-value problem in which the initial conditions are incorporated into the field equations. Thus, $S = \{\mu_i, \varphi_i, e_{ij}, \varkappa_{ij}, \varkappa_{ijk}, \tau_{ij}, m_{ij}, \mu_{ijk}\}$ is a solution of the boundary-initial value problem if and only if S satisfies the equations (4.6), (2.11), (2.17) and the boundary conditions (2.22).

Let us consider the body subjected to two different systems of loadings $L^{(\alpha)} = \{F_i^{(\alpha)}, M_i^{(\alpha)}, F_{ij}^{(\alpha)}, \tilde{u}_i^{(\alpha)}, \tilde{P}_i^{(\alpha)}, \tilde{\varphi}_i^{(\alpha)}, \tilde{m}_i^{(\alpha)}, g_i^{(\alpha)}, \tilde{R}_i^{(\alpha)}, \tilde{Q}_i^{(\alpha)}, a_i^{(\alpha)}, b_i^{(\alpha)}, c_i^{(\alpha)}, d_i^{(\alpha)}\}$, $(\alpha = 1, 2)$ and the two corresponding states $C^{(\alpha)} = \{u_i^{(\alpha)}, \varphi_i^{(\alpha)}, e_{ij}^{(\alpha)}, \varkappa_{ij}^{(\alpha)}, \varkappa_{ijk}^{(\alpha)}, \tau_{ij}^{(\alpha)}, m_{ij}^{(\alpha)}, \mu_{ijk}^{(\alpha)}\}$. We introduce the notations

$$f_i^{(\alpha)} = g * (F_i^{(\alpha)} - F_{ji,j}^{(\alpha)}) + \varrho(tb_i^{(\alpha)} + a_i^{(\alpha)}) \,, \qquad l_i^{(\alpha)} = g * M_i^{(\alpha)} + I_{ij}(td_j^{(\alpha)} + c_j^{(\alpha)}) \,.$$

Reciprocal theorem. *If an elastic solid is subjected to two systems of loadings $L^{(\alpha)}$ $(\alpha = 1, 2)$, then between the corresponding states $C^{(\alpha)}$ there is the following relation*

$$\int_B [f_i^{(1)} * u_i^{(2)} + l_i^{(1)} * \varphi_i^{(2)}] \, dV + \int_{\partial B} g * [(P_i^{(1)} + F_{ji}^{(1)} n_j) * u_i^{(2)} + m_i^{(1)} * \varphi_i^{(2)} + R_i^{(1)} * Du_i^{(2)}] \, dA$$
$$+ \int_\Gamma g * Q_i^{(1)} * u_i^{(2)} \, ds = \int_B [f_i^{(2)} * u_i^{(1)} + l_i^{(2)} * \varphi_i^{(1)}] \, dV + \int_{\partial B} g * [(P_i^{(2)} + F_{ji}^{(2)} n_j) * u_i^{(1)} + m_i^{(2)} * \varphi_i^{(1)}$$
$$+ R_i^{(2)} * Du_i^{(1)}] \, dA + \int_\Gamma g * Q_i^{(2)} * u_i^{(1)} \, ds \,. \tag{4.7}$$

Proof. We will use the method given in [19], [20]. On the basis of the relations (2.16), we conclude from the constitutive equations that

$$\tau_{ij}^{(1)} * e_{ij}^{(2)} + m_{ij}^{(1)} * \varkappa_{ij}^{(2)} + \mu_{ijk}^{(1)} * \varkappa_{ijk}^{(2)} = \tau_{ij}^{(2)} * e_{ij}^{(1)} + m_{ij}^{(2)} * \varkappa_{ij}^{(1)} + \mu_{ijk}^{(2)} * \varkappa_{ijk}^{(1)} \,. \tag{4.8}$$

If we introduce the notations

$$L_{\alpha\beta} = \int_B g * (\tau_{ij}^{(\alpha)} * e_{ij}^{(\beta)} + m_{ij}^{(\alpha)} * \varkappa_{ij}^{(\beta)} + \mu_{ijk}^{(\alpha)} * \varkappa_{ijk}^{(\beta)}) \, dV \,, \tag{4.9}$$

then from (4.8) we have

$$L_{12} = L_{21} \,. \tag{4.10}$$

Using the equations (2.11), (4.6) and the divergence theorem we get

$$L_{\alpha\beta} = \int_{\partial B} g * [(\tau_{ji}^{(\alpha)} - \mu_{sji,s}^{(\alpha)}) n_j * u_i^{(\beta)} + m_i^{(\alpha)} * \varphi_i^{(\beta)} + \mu_{sji}^{(\alpha)} n_s * u_{i,j}^{(\beta)}] \, dA +$$
$$+ \int_B [f_i^{(\alpha)} * u_i^{(\beta)} + l_i^{(\alpha)} * \varphi_i^{(\beta)}] \, dV - \int_B [\varrho u_i^{(\alpha)} * u_i^{(\beta)} + I_{ij} \varphi_i^{(\alpha)} * \varphi_j^{(\beta)}] \, dV$$
$$= \int_{\partial B} g * [(P_i^{(\alpha)} + F_{ji}^{(\alpha)} n_j) * u_i^{(\beta)} + m_i^{(\alpha)} * \varphi_i^{(\beta)} + R_i^{(\alpha)} * Du_i^{(\beta)}] \, dA + \int_\Gamma g * Q_i^{(\alpha)} * u_i^{(\beta)} \, ds +$$
$$+ \int_B [f_i^{(\alpha)} * u_i^{(\beta)} + l_i^{(\alpha)} * \varphi_i^{(\beta)}] \, dV - \int_B [\varrho u_i^{(\alpha)} * u_i^{(\beta)} + I_{ij} \varphi_i^{(\alpha)} * \varphi_j^{(\beta)}] \, dV \,.$$

From (4.10) and (4.11) we obtain (4.7).

By using (2.11) and (2.17), the equations (4.6) can be expressed in terms of displacements and microrotations. The resulting equations can be written in the form

$$\boldsymbol{L}\boldsymbol{u} = \boldsymbol{F}\,, \tag{4.12}$$

where $\boldsymbol{u} = (u_i, \varphi_i)$, $\boldsymbol{F} = (f_i, l_i)$ and the components of the operator $\boldsymbol{L}$ can be found easily.

In what follows we consider the case of homogeneous boundary conditions. With the notations (3.11), the relation (4.7) reduces to

$$\int_B \boldsymbol{v} * \boldsymbol{L}\boldsymbol{u}\,\mathrm{d}V = \int_B \boldsymbol{u} * \boldsymbol{L}\boldsymbol{v}\,\mathrm{d}V\,. \tag{4.13}$$

This relation shows that for homogeneous boundary conditions the operator $\boldsymbol{L}$ is symmetric in convolution. Let us denote by $\mathcal{D}_{\boldsymbol{L}}$ the domain of the definition of the operator $\boldsymbol{L}$. Following [21], [22] we have the

Variational theorem. *Let $M \subset \mathcal{D}_{\boldsymbol{L}}$ be the set of all admissible vectors $\boldsymbol{u}$ which satisfy the homogeneous boundary conditions, and for each $t \in [0, t_0)$ define the functional $\Lambda_t\{\cdot\}$ on M by*

$$\Lambda_t\{\boldsymbol{u}\} = \int_B [\boldsymbol{u} * \boldsymbol{L}\boldsymbol{u} - 2\boldsymbol{u} * \boldsymbol{F}]\,(x, t)\,\mathrm{d}V_x\,,$$

for every $\boldsymbol{u} \in M$. Then

$$\delta\Lambda_t\{\boldsymbol{u}\} = 0\,, \qquad t \in [0, t_0)\,,$$

at $\boldsymbol{u} \in M$ if and only if $\boldsymbol{u}$ is a solution of the boundary-initial-value problem with homogeneous boundary conditions.

With the help of (2.17), (4.9) and (4.11) we can express the functional $\Lambda_t\{\cdot\}$ in the form

$$\begin{aligned}\Lambda_t\{\boldsymbol{u}\} = \int_B \{g * [&A_{ijrs}e_{ij}(\boldsymbol{u}) * e_{rs}(\boldsymbol{u}) + 2B_{ijrs}e_{ij}(\boldsymbol{u}) * \varkappa_{rs}(\boldsymbol{u}) + C_{ijrs}\varkappa_{ij}(\boldsymbol{u}) * \varkappa_{rs}(\boldsymbol{u}) + 2D_{ijpqr}e_{ij}(\boldsymbol{u}) * \varkappa_{pqr}(\boldsymbol{u}) +\\ &+ 2E_{ijpqr}\varkappa_{ij}(\boldsymbol{u}) * \varkappa_{pqr}(\boldsymbol{u}) + F_{ijkpqr}\varkappa_{ijk}(\boldsymbol{u}) * \varkappa_{pqr}(\boldsymbol{u})]\}\,(x, t)\,\mathrm{d}V_x +\\ &+ \int_B [\varrho u_i * u_i + I_{ij}\varphi_i * \varphi_j]\,(x, t)\,\mathrm{d}V_x - 2\int_B [f_i * u_i + g_i * \varphi_i]\,(x, t)\,\mathrm{d}V_x\,,\end{aligned}$$

where

$$e_{ij}(\boldsymbol{u}) = u_{j,i} + \varepsilon_{kji}\varphi_k\,, \qquad \varkappa_{ij}(\boldsymbol{u}) = \varphi_{j,i}\,, \qquad \varkappa_{ijk}(\boldsymbol{u}) = u_{k,ij}\,.$$

Other variational theorems, similar to those established in [16], [18], [22] can be derived.

References

1 Eringen, A. C., Theory of micropolar elasticity; Fracture, vol. II, Academic Press, New York 1968.
2 Nowacki, W., Theory of micropolar elasticity, Springer, Wien 1970.
3 Eringen, A.; Kafadar, C. B., Polar field theories; in Continuum Physics (ed. A. C. Eringen), vol. IV, Academic Press, New York 1976.
4 Brulin, O., Linear micropolar media, CISM Courses and Lectures **1979.**
5 Hjalmars, S., Nonlinear micropolar theory, CISM Courses and Lectures 1979.
6 Berglund, K., Structural models of micropolar media, CISM Courses and Lectures 1979.
7 Brulin, O.; Hjalmars, S., Grade consistent micropolar theory and equivalent media, TRITA-MEK-79-15, Technical Reports from the Royal Institute of Technology, Dept. of Mechanics, Stockholm, Sweden.
8 Rymarz, Cz., On the model of a non-simple medium with rotational degrees of freedom, Bull. Acad. Polon. Sci., Sér. Sci. Techn. **16**, 271 (1968).
9 Green, A. E.; Rivlin, R. S., Simple force and stress multipoles, Arch. Rational Mech. Anal. **16**, 325 (1964).
10 Toupin, R. A., Theories of elasticity with couple stresses, Arch. Rational Mech. Anal. **17**, 85 (1964).
11 Mindlin, R. D., Microstructure in linear elasticity, Arch. Rational Mech. Anal. **16**, 51 (1964).
12 Hlaváček, I.; Hlaváček, M., On the existence and uniqueness of solution and some variational principles in linear theories of elasticity with couple-stresses, Aplikace Mat. **14**, 411 (1969).
13 Fichera, G., Existence theorems in elasticity, in Encyclopedia of Physics, Vol. VIa/2, Springer, Berlin 1972.
14 Ieşan, D., On the positive definiteness of the operator of micropolar elasticity, J. Eng. Math. 8, 107 (1974).
15 Ieşan, D., On the second boundary value problem in the linear theory of micropolar elasticity, Rend. Accad. Lincei **55**, 456 (1973).
16 Gurtin, M. E., The linear theory of elasticity, in Enciclopedia of Physics (ed. C. Truesdell) vol. VIa/2, Springer, Berlin 1972.
17 Ignaczak, I., A completness problem for stress equation of motion in the linear elasticity theory, Arch. Mech. Stos. **15**, 225 (1963).
18 Ignaczak, J., Tensorial equations of motion for elastic materials with microstructure, in Trends in elasticity and thermoelasticity, Witold Nowacki Anniversary Volume, Wolters-Noordhoff Publ., Groningen 1971.
19 Ieşan, D., Sur la théorie de la thérmoelasticité micropolaire coupleé. C. R. Acad. Sc. Paris, 265 A, 271 (1967).
20 Ieşan, D., On the linear theory of micropolar elasticity, Int. J. Eng. Sci., **7**, 1213 (1969).
21 Hlaváček, I., Variational formulation of the Cauchy problem for equations with operator coefficients, Aplikace matematiky **16**, 46 (1971).
22 Ieşan, D., On reciprocity theorems and variational theorems in linear elastodynamics, Bull. Acad. Polon. Sci., Sér. Sci. Techn. **22**, 273 (1974).

Eingereicht am 25. 3. 81

Anschrift: Prof. D. Ieşan, Faculty of Mathematics, University of Iaşi, Iaşi, Romania

ZAMM 62, 43–50 (1982)

R. K. JAIN / SAROJ PRABHA

Gas-Particulate Flow through a Curved Pipe

Eine Gas-Feststoffströmung durch ein gekrümmtes Rohr wird mit dem Ziel der Bestimmung des Einflusses der Sekundärströmung auf den Volumenstrom bei einem gegebenen Druckgradienten untersucht. Es wird festgestellt, daß mit zunehmender Sekundärströmung der Volumenstrom sinkt. Bei kleinen Teilchen ist darüberhinaus der Volumenstrom für das reine Gas größer als der Volumenstrom der Gas-Feststoffströmung. Bei groben Teilchen ist der Volumenstrom der Gas-Feststoffströmung größer als der Volumenstrom für das reine Gas.

Gas-particulate flow through the curved pipe is analyzed in order to study the effect of the secondary flow on the volume flow rate at given pressure gradient. It is observed that as the secondary flow through the pipe increases, volume flow rate decreases. Moreover, when the particles are fine, the volume flow rate for the clean gas is greater than the volume flow rate for gas-particulate flow. For coarse particles, volume flow rate for gas-particulate flow is greater than the volume flow rate for the clean gas.

Рассматривается течение газа вместе с твёрдыми частицами по искривленной трубе, чтобы определить влияние вторичного течения на расход при заданном градиенте давления. Оказывается, что с ростом вторичного течения расход уменьшается. Более того, для маленьких частиц расход чистого газа больше расхода смеси ,,газ-частицы‘‘. Для более крупных частиц расход смеси ,,газ-частицы‘‘ больше расхода чистого газа.

1. Introduction

In a cross-section of a curved pipe, under the action of centrifugal forces and pressure gradient, a secondary motion is set up which greatly reduces the volume flow rate. Hence to get a certain volume flow rate, a larger pressure gradient is required in a curved pipe than it is needed in the straight one. One may be referred to the works of EUSTICE [4], DEAN [1, 2, 3] and SMITH [6] for theoretical and experimental works on secondary motion of one phase flow in a curved pipe. These investigations are based on the assumption that the radius of curvature of the pipe is large in comparison to the radius of the cross-section.

To the author's knowledge, no attempt has so far been made to study the nature of the two phase flow through a curved pipe. In the present analysis, we extend the work of DEAN [3] to gas-particulate flow through a curved pipe for studying the effect of secondary motion on the volume flow rate at a given pressure gradient. We assume that (i) the radius of the circle in which the central line of the pipe is coiled is large in comparison with the radius of the cross-section of the pipe. (ii) two phase flow is a slight variant of the one phase flow and therefore the actual secondary motion of the gas phase is a uniform stream in the central region. We calculate the stream function for the particle phase from the governing equations. Green's function approach is used to obtain the closed form solutions for the pipe of a rectangular and circular cross-sections. DEAN's [3] case comes out as a particular case for the clean gas.

We find that for a given pressure gradient, the secondary flow effects the volume flow rate in the presence of the particles also. When the particles are fine in nature, the volume flow rate for the clean gas is more than the volume flow rate for the gas-particulate flow. For coarse particles, the volume flow rate for gas-particulate system is more than the volume flow rate for the clean gas flow. In both the cases, it is obtained that the region where the primary motion is greatest shifts outward.

2. Mathematical Formulation of the Problem

Consider the steady gas-particulate flow through an infinitely long curved pipe. The radius of curvature R of the pipe at any point is large in comparison to the radius of the cross-section a. Fig. 1a shows the system of coordinates for considering the flow through a curved pipe. OZ is the axis of the anchor ring formed by the pipe wall, C is the centre of the cross-section of the pipe by a plane through OZ that makes an angle φ with a fixed plane and CO is perpendicular to OZ and is of length R. If P is any point in the cross-section and the distance of the point P from OZ is r, then the position of the point P is specified by cylindrical coordinates (r, φ, z). The surface of the pipe is given by $r = a$, a being the radius of any section.

Let the components of velocities of the gas and particle phases be denoted by (U, V, W) and (U_p, V_p, W_p) respectively. Let ν, ϱ and P represent the kinematic viscosity, density and pressure for the gas phase and ϱ_p, ϱ_{sp} and d the particulate density, material density and diameter of the spherical particles respectively. If the density ϱ and ϱ_p of the ambient fluid and that of the particle phase are assumed to be constant, then the basic equations for two phases in cylindrical polar coordinates can be reduced to the following:

$$\nu\left(\frac{\partial^2}{\partial r^2}+\frac{\partial^2}{\partial z^2}\right)V = -\frac{A}{\varrho}+\left(\frac{\partial\psi}{\partial r}\frac{\partial}{\partial z}-\frac{\partial\psi}{\partial z}\frac{\partial}{\partial r}\right)V+\frac{F\varrho_p}{\varrho}(V-V_p)\,, \tag{2.1}$$

$$\nu\left(\frac{\partial^2}{\partial r^2}+\frac{\partial^2}{\partial z^2}\right)^2\psi = \left(\frac{\partial\psi}{\partial r}\frac{\partial}{\partial z}-\frac{\partial\psi}{\partial z}\frac{\partial}{\partial r}\right)\left(\frac{\partial^2\psi}{\partial z^2}+\frac{\partial^2\psi}{\partial r^2}\right)+\frac{2V}{R}\frac{\partial V}{\partial z}+\frac{F\varrho_p}{\varrho}\left(\frac{\partial^2\psi}{\partial z^2}+\frac{\partial^2\psi}{\partial r^2}\right)-\frac{F\varrho_p}{\varrho}\left(\frac{\partial^2\psi_p}{\partial z^2}+\frac{\partial^2\psi_p}{\partial r^2}\right), \tag{2.2}$$

$$\left(\frac{\partial\psi_p}{\partial r}\frac{\partial}{\partial z} - \frac{\partial\psi_p}{\partial z}\frac{\partial}{\partial r}\right)\frac{\partial\psi_p}{\partial z} + \frac{V_p^2}{R} = F\left(\frac{\partial\psi}{\partial z} - \frac{\partial\psi_p}{\partial z}\right), \tag{2.3}$$

$$\left(\frac{\partial\psi_p}{\partial r}\frac{\partial}{\partial z} - \frac{\partial\psi_p}{\partial z}\frac{\partial}{\partial r}\right)V_p = F(V - V_p), \qquad \left(\frac{\partial\psi_p}{\partial r}\frac{\partial}{\partial z} - \frac{\partial\psi_p}{\partial z}\frac{\partial}{\partial r}\right)\frac{\partial\psi_p}{\partial r} = F\left(\frac{\partial\psi}{\partial r} - \frac{\partial\psi_p}{\partial r}\right) \qquad (2.4), (2.5)$$

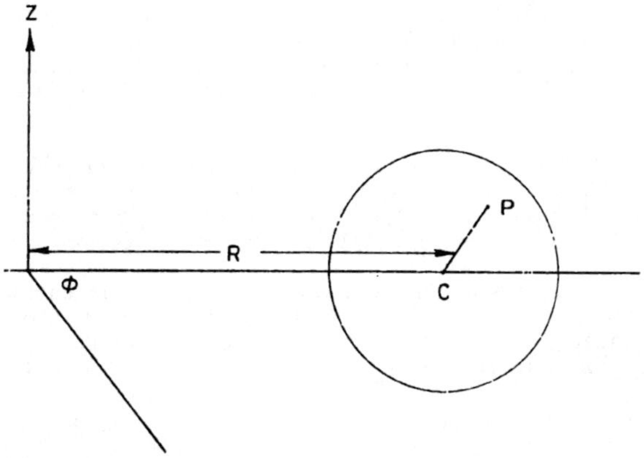

Fig 1a. Coordinate system

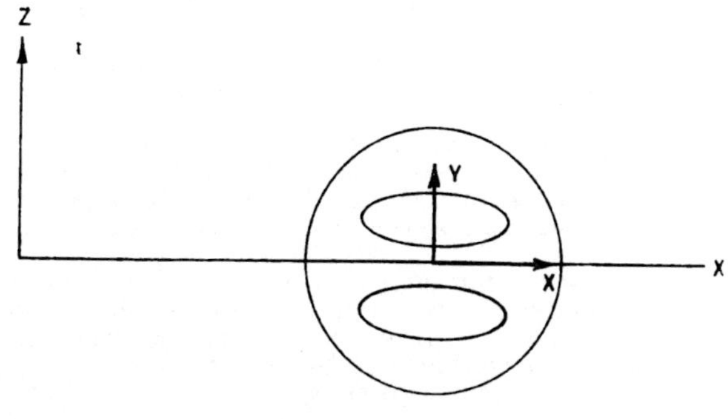

Fig 1b. Transformed coordinates

where ψ and ψ_p denote stream functions for the gas and particle phases respectively such that

$$U = -\frac{\partial\psi}{\partial z}, \qquad W = \frac{\partial\psi}{\partial r}, \qquad U_p = -\frac{\partial\psi_p}{\partial z}, \qquad \text{and} \qquad W_p = \frac{\partial\psi_p}{\partial r}. \tag{2.6}$$

In deriving these equations, the radius of curvature R of the pipe is assumed to be large as compared to the radius a of the cross-section and as a consequence $\frac{1}{r}$ and $\frac{\partial}{\partial r} + \frac{1}{r}$ are approximated by $\frac{1}{R}$ and $\frac{\partial}{\partial r}$ respectively. All the flow variables except the pressure are assumed to be independent of φ. This and the assumption that $R \gg a$ leads to the consequence that space rate of decrease in pressure along the central line is a constant. It is denoted by A in the basic equations. F denotes the interaction parameter given by

$$F = \frac{18\nu\varrho}{d^2\varrho_{sp}}.$$

In the approximate eqs. (2.1) to (2.5), the coordinates r and z appear only in the operators $\frac{\partial}{\partial r}$ and $\frac{\partial}{\partial z}$, hence any point in the plane that is selected for reference can be taken as the origin and we denote X, Y for the corresponding coordinates [Fig. 1b]. We non-dimensionalize the variables; current coordinates X, Y with respect to a, velocities with V_0 $(= \varrho A a^2/4\nu)$, stream functions with ν. If $x, y, v, v_p, \Phi, \Phi_p$ and $F'(= Fa^2/\nu)$ are non-dimensional forms of $X, Y, V, V_p, \psi, \psi_p$ and F respectively, then the equations (2.1) to (2.5) reduce to the form

$$\left(\frac{\partial^2}{\partial x^2} + \frac{\partial^2}{\partial y^2}\right)v = -4 + \left(\frac{\partial\Phi}{\partial x}\frac{\partial}{\partial y} - \frac{\partial\Phi}{\partial y}\frac{\partial}{\partial x}\right)v + \sigma_p\left(\frac{\partial\Phi_p}{\partial x}\frac{\partial}{\partial y} - \frac{\partial\Phi_p}{\partial y}\frac{\partial}{\partial x}\right)v_p, \tag{2.7}$$

$$\left(\frac{\partial^2}{\partial x^2} + \frac{\partial^2}{\partial y^2}\right)^2\Phi = \left(\frac{\partial\Phi}{\partial x}\frac{\partial}{\partial y} - \frac{\partial\Phi}{\partial y}\frac{\partial}{\partial x}\right)\left(\frac{\partial^2\Phi}{\partial x^2} + \frac{\partial^2\Phi}{\partial y^2}\right) + Kv\frac{\partial v}{\partial y} + F'\sigma_p\left(\frac{\partial^2\Phi}{\partial x^2} + \frac{\partial^2\Phi}{\partial y^2}\right) - F'\sigma_p\left(\frac{\partial^2\Phi_p}{\partial x^2} + \frac{\partial^2\Phi_p}{\partial y^2}\right), \tag{2.8}$$

$$\left(\frac{\partial\Phi_p}{\partial x}\frac{\partial}{\partial y} - \frac{\partial\Phi_p}{\partial y}\frac{\partial}{\partial x}\right)\frac{\partial\Phi_p}{\partial y} + \frac{K}{2}v_p^2 = F'\left(\frac{\partial\Phi}{\partial y} - \frac{\partial\Phi_p}{\partial y}\right), \tag{2.9}$$

$$\left(\frac{\partial\Phi_p}{\partial x}\frac{\partial}{\partial y} - \frac{\partial\Phi_p}{\partial y}\frac{\partial}{\partial x}\right)\frac{\partial\Phi_p}{\partial x} = F'\left(\frac{\partial\Phi}{\partial x} - \frac{\partial\Phi_p}{\partial x}\right) \tag{2.10}$$

where V_0 is the velocity for the gas phase on the axis of the circular pipe and

$$K = \frac{2a}{R}\left(\frac{V_0 a}{\nu}\right)^2, \tag{2.11}$$

K is the measure of the curvature of the pipe and it depends on flow Reynolds number $Re\left(= \frac{V_0 a}{\nu}\right)$. K is zero for a straight pipe.

The structure of the basic equations is not significantly altered by the presence of particulate phase. For a given velocity distribution of particulate phase, equation (2.7) again represents deflection of a flexible membrane due to pressure. It is therefore, natural to believe, in view of the aimes of the present study that the gas phase flow will qualitatively be identical to that proposed by DEAN [3]. In the central portion of a cross-section of the tube, secondary flow for the gas phase will be assumed to be that of the streaming in nature along X, giving there by

$$\Phi = -2ky, \qquad k \text{ being a positive integer} \tag{2.12}$$

Knowledge of Φ helps to determine Φ_p from equation (2.10). Since the stream lines of particulate phase, in central region, are regarded as slight variant of the ones of gas phase, we seek a separable solution of equation (2.10) for Φ_p with Φ as given by (2.12). We obtain

$$\Phi_p = \frac{1}{\lambda} y(F' + \lambda C\, e^{\lambda x}) , \tag{2.13}$$

C and λ are certain constants of integration.

We know that the smaller is the size of the dust particle, the smaller will be the relaxation time implying thereby that fine dust adjusts more quickly to the surrounding fields as compared to coarse dust. It is also clear that F' is large for fine dust and small for coarse dust as it is inversely proportional to d^2. In equation (2.13), therefore $\lambda C\, e^{\lambda x}$ could be neglected (retained) as compared to F' for fine (coarse) dust in the central region. We will, therefore, throughout regard

$$\begin{aligned} \Phi_p &= \frac{1}{\lambda} yF' \quad \text{fine dust}, \\ \Phi_p &= Cy\, e^{\lambda x} \quad \text{coarse dust}. \end{aligned} \tag{2.14 a, b}$$

One observes that Φ_p, for λ negative, is of the same nature as Φ. Secondary flow for fine particles has, therefore, been approximated by a uniform stream. To have an analogy with Φ, C and λ will be regarded as negative real numbers.

Using eqs. (2.12)—(2.13), eq. (2.9) gives

$$v_p = \left[\frac{2C^2\lambda\, e^{2\lambda x}}{K} - \frac{4kF'}{K} - \frac{2F'^2}{K\lambda'}\right]^{1/2}. \tag{2.15}$$

On making use of eqs. (2.12), (2.13) and (2.15), eq. (2.7) gives

$$\frac{\partial^2 v}{\partial x^2} + \frac{\partial^2 v}{\partial y^2} - 2k\frac{\partial v}{\partial x} = -4 - \frac{2C^2\lambda^2\left(\dfrac{F'}{\lambda} + C\, e^{\lambda x}\right) e^{2\lambda x}\sigma_p}{K^{1/2}[2C^2\lambda\, e^{2\lambda x} - 4kF' - 2F'^2/\lambda]^{1/2}}. \tag{2.16}$$

This equation will be solved for the pipe of a rectangular cross-section with the boundary conditions that v vanishes on the surface. Due to the presence of the second term on the RHS of eq. (2.16), the method of separation of variables as adopted by DEAN [3] fails to give a closed form solution. As such, we solve the boundary value problem so obtained by the method of GREEN's function.

3. Solution of the Problem

Let the pipe be of rectangular cross-section given by $x = \pm 1$ and $y = \pm b$.

Solution of eq. (2.16) subject to the boundary conditions

$$v = 0 \quad \text{at} \quad x = \pm 1 \quad \text{and} \quad y = \pm b \tag{3.1}$$

is

$$v(x, y) = \int_{-1}^{1}\int_{-b}^{b}\left[4 + \frac{2C^2\lambda^2\left(\dfrac{F'}{\lambda} + C\, e^{\lambda\xi}\right) e^{2\lambda\xi}\sigma_p}{K^{1/2}[2C^2\lambda\, e^{2\lambda\xi} - 4kF' - 2F'^2/\lambda]^{1/2}}\right] G(x, y; \xi, \eta)\, d\xi\, d\eta \tag{3.2}$$

where $G(x, y; \xi, \eta)$ is Green's function for the homogeneous boundary value problem associated with (2.16) and (3.1):

$$\begin{aligned} G(x, y; \xi, \eta) &= -\sum_{m=0}^{\infty} \frac{\cos(\mu_m\eta)\cos(\mu_m y)}{b\sigma_m \sinh 2\sigma_m} e^{-k(\xi - x)} \sinh\sigma_m(\xi - 1)\sinh\sigma_m(x + 1), \qquad -1 \leqq x < \xi, \\ G(x, y; \xi, \eta) &= -\sum_{m=0}^{\infty} \frac{\cos(\mu_m\eta)\cos(\mu_m y)}{b\sigma_m \sinh 2\sigma_m} e^{-k(\xi - x)} \sinh\sigma_m(\xi + 1)\sinh\sigma_m(x - 1), \qquad \xi < x \leqq 1 \end{aligned} \tag{3.3}$$

where

$$\mu_m = \cos\left(m + \frac{1}{2}\right)\frac{\pi}{b}, \qquad \sigma_m = \sqrt{k^2 + \mu_m^2}.$$

The solution corresponding to first term in the integrand in eq. (3.2) represents the case of clean gas. Using the identities

$$\text{(i)} \quad \int e^{ax} \sinh bx \, dx = \frac{e^{ax}}{a^2 - b^2}(a \sinh bx - b \cosh bx), \tag{3.4}$$

$$\text{(ii)} \quad y^2 - b^2 = \sum_{m=0}^{\infty} \frac{4(-1)^m \cos(\mu_m y)}{b\mu_m^3} \tag{3.5}$$

the solution corresponding to the clean gas is

$$2(b^2 - y^2) + 8\,e^{kx} \sum_{m=0}^{\infty} \left[\frac{(-1)^m \sinh k \sinh \sigma_m x \cos \mu_m y}{b\mu_m^3 \sinh \sigma_m} + \frac{(-1)^{m+1} \cosh k \cosh \sigma_m x \cos \mu_m y}{b\mu_m^3 \cosh \sigma_m}\right]. \tag{3.6}$$

This solution is the same as obtained by DEAN [3] by the method of separation of variables.

The solution corresponding to the second term in the integrand in eq. (3.3) represents the case of particulate flow. Here, we integrate this term in the following cases:

Case I: Fine dust ($F' \gg \lambda C\,e^{\lambda x}$)

The second term within the bracket of the integrand of (3.2) reduces to

$$\frac{2C^2\lambda^2\left(\frac{F'}{\lambda} + C\,e^{\lambda\xi}\right) e^{2\lambda\xi}\,\sigma_p}{K^{1/2}[-2F'(2k + F'/\lambda)]^{1/2}};$$

with this simplification, the solution corresponding to the particulate flow is

$$-\sum_{m=0}^{\infty} \frac{4(-1)^m DF'\sigma_p C^2\lambda\,e^{kx} \cos \mu_m y}{b\mu_m H[(2\lambda - k)^2 - \sigma_m^2] \sinh 2\sigma_m} - \sum_{m=0}^{\infty} \frac{4(-1)^m \sigma_p E C^3\lambda^2\,e^{kx} \cos \mu_m y}{b\mu_m H[(3\lambda - k)^2 - \sigma_m^2] \sinh 2\sigma_m} -$$

$$-\sum_{m=0}^{\infty} \frac{4(-1)^m \sigma_p C^2\lambda F'\,e^{2\lambda x} \cos \mu_m y}{b\mu_m H[(2\lambda - k)^2 - \sigma_m^2]} - \sum_{m=0}^{\infty} \frac{4(-1)^m \sigma_p C^3\lambda^2\,e^{3\lambda x} \cos \mu_m y}{b\mu_m H[(3\lambda - k)^2 - \sigma_m^2]} \tag{3.7}$$

where

$$D = e^{-(2\lambda - k)} \sinh \sigma_m(x - 1) - e^{(2\lambda - k)} \sinh \sigma_m(x + 1),$$

$$E = e^{-(3\lambda - k)} \sinh \sigma_m(x - 1) - e^{(3\lambda - k)} \sinh \sigma_m(x + 1).$$

The complete solution $v(x, y)$ of eq. (3.2) is now the sum of the expressions (3.6) and (3.7):

$$\text{Volume rate of Flow} = F_1(\text{say}) = \int_{-1}^{1}\int_{-b}^{b} v(x, y)\,dx\,dy =$$

$$= \frac{16b^3}{3} - \frac{16}{b}\sum_{m=0}^{\infty} \frac{\sigma_m(\cosh 2\sigma_m - \cosh 2k)}{\mu_m^6 \sinh \sigma_m \cosh \sigma_m} - \sum_{m=0}^{\infty} \frac{16F'C^2\lambda L\sigma_p}{b\mu_m^4 H[(2\lambda - k)^2 - \sigma_m^2] \sinh 2\sigma_m}$$

$$- \sum_{m=0}^{\infty} \frac{16C^3\lambda^2 M\sigma_p}{b\mu_m^4 H[(3\lambda - k)^2 - \sigma_m^2] \sinh 2\sigma_m} - \sum_{m=0}^{\infty} \frac{8\sigma_p C^2\lambda F' \sinh 2\lambda}{b\mu_m^2\lambda H[(2\lambda - k)^2 - \sigma_m^2]}$$

$$- \sum_{m=0}^{\infty} \frac{16C^3\lambda\sigma_p \sinh 3\lambda}{3b\mu_m^2 H[(3\lambda - k)^2 - \sigma_m^2]} \tag{3.8}$$

where

$$L = k \sinh 2\sigma_m \sinh 2\lambda - \sigma_m \cosh 2\sigma_m \cosh 2\lambda + \sigma_m \cosh(2\lambda - 2k),$$

$$M = k \sinh 2\sigma_m \sinh 3\lambda - \sigma_m \cosh 2\sigma_m \cosh 3\lambda + \sigma_m \cosh(3\lambda - 2k).$$

The first two terms of (3.8) give the case of particle free fluid which agrees with DEAN's [3] results.

Case II: Coarse Particles ($F' \ll \lambda C\,e^{\lambda x}$)

The term related to particle phase in the integrand of (3.2) reduces to

$$\sigma_p \sqrt{\frac{2\lambda^3}{K}}\,C\,e^{\lambda\xi}\left(\frac{F'}{\lambda} + C\,e^{\lambda\xi}\right) + \sigma_p \sqrt{\lambda^3/2K}\,C\left(\frac{F'^2}{C^2\lambda^2} + \frac{2kF'}{C^2\lambda}\right)\left(\frac{F'}{\lambda}\,e^{-\lambda\xi} + C\right);$$

with this simplification, the solution corresponding to the particulate flow is

$$
\begin{aligned}
&-\sum_{m=0}^{\infty} \frac{2(-1)^m F'\sigma_p C\sqrt{2\lambda^3/K}\, e^{kx}\cos\mu_m y}{b\lambda\mu_m[(\lambda-k)^2-\sigma_m^2]\sinh 2\sigma_m}\,[e^{-(\lambda-k)}\sinh\sigma_m(x-1)-e^{(\lambda-k)}\sinh\sigma_m(x+1)] - \\
&-\sum_{m=0}^{\infty} \frac{2(-1)^m CF'\sigma_p\sqrt{2\lambda^3/K}\, e^{\lambda x}\cos\mu_m y}{b\lambda\mu_m[(\lambda-k)^2-\sigma_m^2]} - \\
&-\sum_{m=0}^{\infty} \frac{2(-1)^m C^2\sqrt{2\lambda^3/K}\, e^{kx}\sigma_p\cos\mu_m y}{b\mu_m[(2\lambda-k)^2-\sigma_m^2]\sinh 2\sigma_m}\,[e^{-(2\lambda-k)}\sinh\sigma_m(x-1)-e^{(2\lambda-k)}\sinh\sigma_m(x+1)] - \\
&-\sum_{m=0}^{\infty} \frac{2(-1)^m\sqrt{2\lambda^3/K}\, C^2 e^{2\lambda x}\sigma_p\cos\mu_m y}{b\mu_m[(2\lambda-k)^2-\sigma_m^2]} - \\
&-\sum_{m=0}^{\infty} \frac{2(-1)^m F'\sigma_p CP_1\sqrt{\lambda^3/2K}\, e^{kx}\cos\mu_m y}{b\lambda\mu_m[(\lambda+k)^2-\sigma_m^2]\sinh 2\sigma_m}\,[e^{(\lambda+k)}\sinh\sigma_m(x-1)-e^{-(\lambda+k)}\sinh\sigma_m(x+1)] + \\
&+\sum_{m=0}^{\infty} \frac{2(-1)^m C^2P_1\sqrt{\lambda^3/2K}\, e^{kx}\sigma_p\cos\mu_m y}{b\mu_m^3\sinh 2\sigma_m}\,[e^{k}\sinh\sigma_m(x-1)-e^{-k}\sinh\sigma_m(x+1)] - \\
&-\sum_{m=0}^{\infty} \frac{2(-1)^m \sigma_p CP_1\sqrt{\lambda^3/2K}\, e^{-\lambda x}\cos\mu_m y}{b\lambda\mu_m[(\lambda+k)^2-\sigma_m^2]} + \sum_{m=0}^{\infty}\frac{2(-1)^m P_1C^3\sqrt{\lambda^3/2K}\,\sigma_p\cos\mu_m y}{b\mu_m^3}
\end{aligned}
\tag{3.9}
$$

where

$$P_1 = \frac{F'^2}{C^2\lambda^2} + \frac{2kF'}{C^2\lambda}.$$

The complete solution $v(x, y)$ of eq. (3.3) in this case is the sum of the expressions (3.6) and (3.9):

$$\text{Volume flow rate} = F_2(\text{say}) = \int_{-1}^{1}\int_{-b}^{b} v(x, y)\,dx\,dy =$$

$$
\begin{aligned}
&= 16\frac{b^3}{3} - \frac{16}{b}\sum_{m=0}^{\infty}\frac{\sigma_m(\cosh 2\sigma_m-\cosh 2k)}{\mu_m^6\sinh\sigma_m\cosh\sigma_m} - \\
&-\sum_{m=0}^{\infty}\frac{8F'\sigma_p C\sqrt{2\lambda^3/K}}{b\mu_m^4\lambda[(\lambda-k)^2-\sigma_m^2]\sinh 2\sigma_m}\,[\sigma_m\cosh(\lambda-2k)+k\sinh 2\sigma_m\sinh\lambda-\sigma_m\cosh 2\sigma_m\cosh\lambda] - \\
&-\sum_{m=0}^{\infty}\frac{8LC^2\sqrt{2\lambda^3/K}\,\sigma_p}{b\mu_m^4\lambda[(2\lambda-k)^2-\sigma_m^2]\sinh 2\sigma_m} - \\
&-\sum_{m=0}^{\infty}\frac{8F'\sigma_p P_1 C\sqrt{2\lambda^3/K}}{b\lambda\mu_m^4[(\lambda+k)^2-\sigma_m^2]\sinh 2\sigma_m}\,[\sigma_m\cosh(\lambda+2k)-k\sinh 2\sigma_m\sinh\lambda-\sigma_m\cosh 2\sigma_m\cosh\lambda] + \\
&+\sum_{m=0}^{\infty}\frac{8C^2P_1\sqrt{2\lambda^3/K}\,(\sigma_m\cosh 2k-\sigma_m\cosh 2\sigma_m)\,\sigma_p}{b\mu_m^6\sinh 2\sigma_m} - \sum_{m=0}^{\infty}\frac{8CF'\sigma_p\sqrt{2\lambda^3/K}\,\sinh\lambda}{b\lambda^2\mu_m^2[(\lambda-k)^2-\sigma_m^2]} - \\
&-\sum_{m=0}^{\infty}\frac{8C^2\sqrt{2\lambda^3/K}\,\sigma_p\sinh\lambda}{2b\lambda\mu_m^2[(2\lambda-k)^2-\sigma_m^2]} + \sum_{m=0}^{\infty}\frac{8C^2P_1\sqrt{\lambda^3/2K}\,\sigma_p}{b\mu_m^4} - \sum_{m=0}^{\infty}\frac{8F'\sigma_p C\sqrt{\lambda^3/2K}\,\sinh\lambda}{b\lambda^2\mu_m^2[(\lambda+k)^2-\sigma_m^2]}.
\end{aligned}
\tag{3.10}
$$

4. Discussion of the Results

In this section, we present certain basic features of gas-particulate flow through a curved pipe. For numerical computation, the following values of the parameters are considered

$$k = 0, 1, 2, \dots, 10; \quad C = -.25, .25; \quad \lambda = -.9, .9,$$

$$F' = .05, 10, 15; \quad K = 1.0, \quad \sigma_p = .25, \quad b = .6, 1.6.$$

Fig. 2 gives the variation of the volume flow rate with respect to the secondary flow parameter k, when the secondary motion for the particle phase is a uniform stream. This figure indicates that as k increases, the volume flow rate decreases. Also, the volume flow rate for the clean gas is greater compared to the flow rate for gas-particulate system. This is due to the fact that particles are fine in this case.

Fig. 3 gives the variation of the volume flow rate with the secondary flow for the coarse particles. This figure also illustrates that the volume flow rate decreases as the secondary flow increases but for the clean gas, volume flow rate is less compared to the flow rate for gas-particulate system.

Figs. 4 and 5 give the variation of velocity v with the secondary flow at $y = 0$ for the cases of fine and coarse dust respectively. Both the figures illustrate that v reduces as the secondary flow through the pipe increases. It is further indicated that the region where the primary motion is greatest, shifts outward. Also, figure 4 shows that v

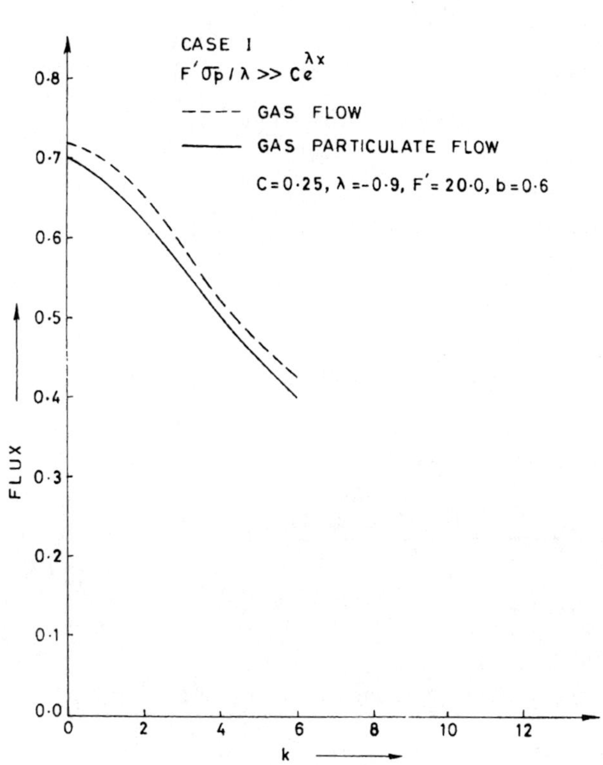

Fig. 2. Variation of volume rate of flow with changes in k

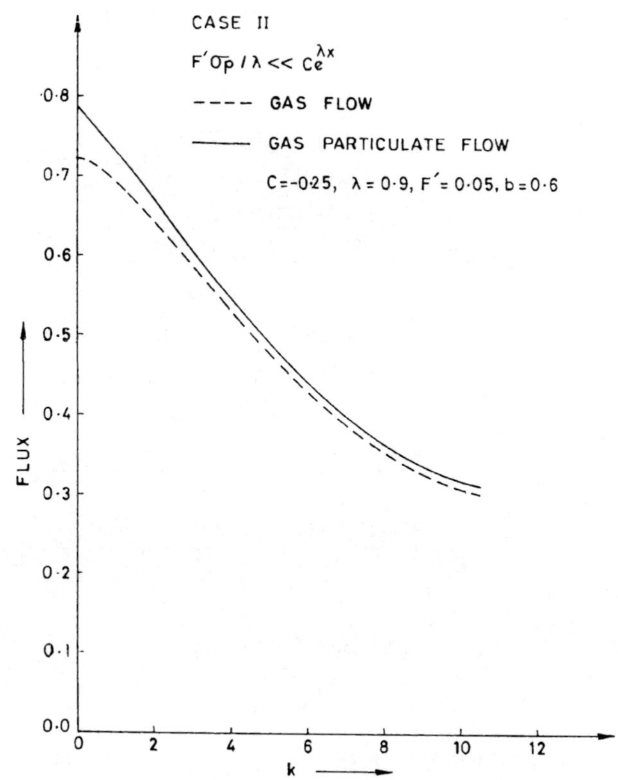

Fig. 3. Variation of volume flow rate with changes in k

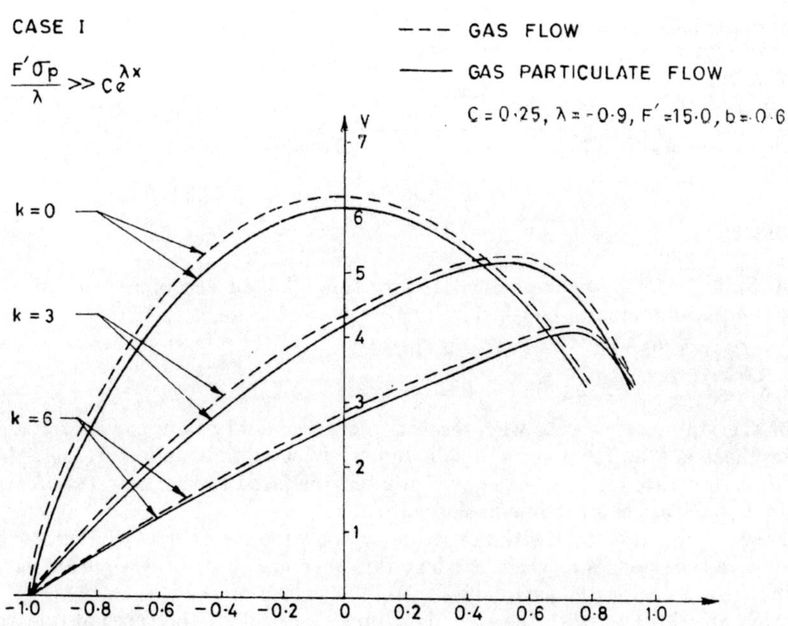

Fig. 4. Variation of V with X at $Y = 0.0$

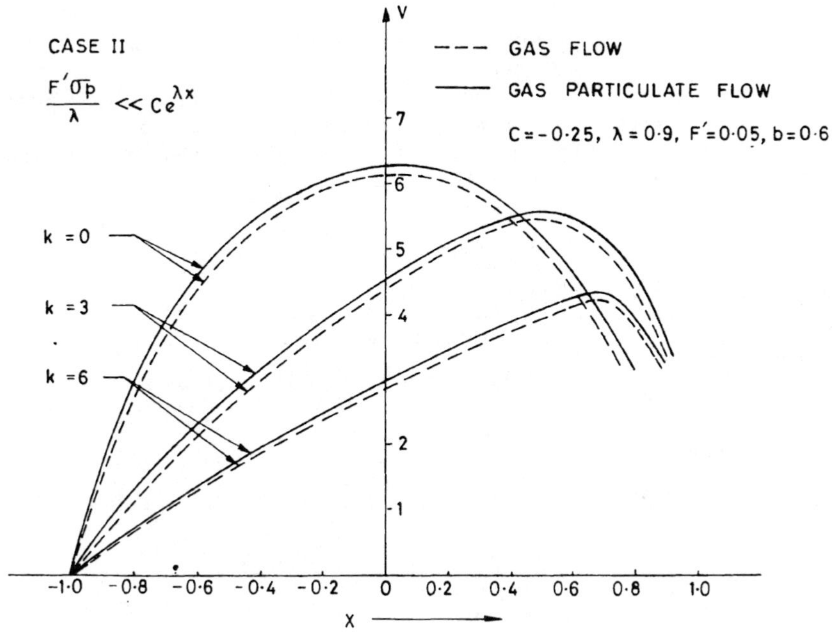

Fig. 5. Variation of V with X at $Y = 0.0$

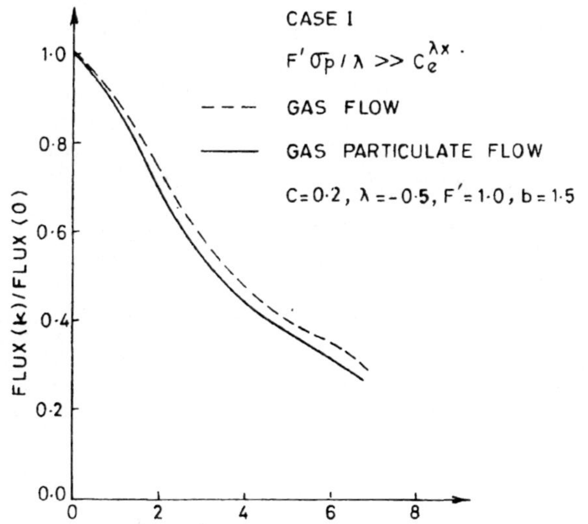

Fig. 6. Variation of volume rate of flow with k

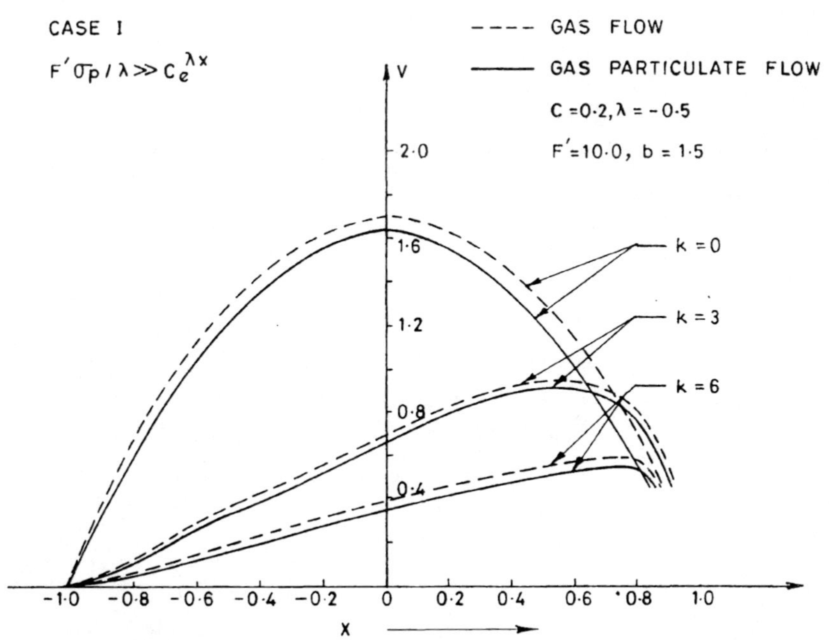

Fig. 7. Variation of V with X at $Y = 0.0$

for the clean gas is greater compared to the v for gas-particulate flow while figure 5 indicates that v for the gas-particulate flow is greater compared to the v for clean gas.

Figs. 6 and 7 give the variation of volume flow rate and velocity v with the secondary flow parameter k for fine dust case. For different set of parameters the same trend is observed as in Figs. 2 and 4.

The details for a pipe of circular cross-section are computationally more tedius and lengthy. Physical behaviour is identical in nature. It has therefore not been included in the present study.

References

1 DEAN, W. R., Note on the Motion of Fluid in a Curved Pipe, Phil. Mag. 4 (1927), p. 208.
2 DEAN, W. R., Stream Line Motion of Fluid in a Curved Pipe, Phil. Mag. 5 (1928), p. 673.
3 DEAN, W. R., Note on the Motion of Fluid in a Curved Pipe, Mathematika, 6 (1959), p. 77.
4 EUSTICE, J., Proc. Roy. Soc. London 85A (1911), p. 122.
5 GRAY, A.; MATHEWS, G. B., A Treatise on Bessel Functions and their Applications to Physics, Second edition, New York, Dover.
6 SMITH, F. T., Steady Motion with in a Curved Pipe, Proc. Roy. Soc. London, 347 A (1976), p. 345.

Eingereicht am 30. 6. 1980, in revidierter Fassung am 2. 3. 1981

Anschrift: Dr. SAROJ PRABHA, Department of Mathematics, Indian Institute of Technology, Kanpur-208016, U.P., India

KLEINE MITTEILUNGEN

ZAMM 62, 51 – 52 (1982)

A. K. Borkakati / I. Pop

On Heat Transfer in Hartmann Flow I

1. Introduction

The flow of fluids in channels is of importance in technological area. Such a study has been undertaken by a number of researchers in the case of flow fields and temperature fields. They have also considered the effects of transversely applied magnetic field on the channel flows which are received considerable attention in recent years due to their practical applications in MHD generators, accelerators etc. A good deal of references of these problems can be found in a recent book by Blum, Michailov and Ozols [1]. But, however, the heat transfer aspects of these electrically conducting fluids in the presence of an applied magnetic field has received less attention. Hence, here some considerations of the heat transfer problem of such a flow are presented.

2. Analysis

We consider the motion of a viscous incompressible electrically conducting liquid between two infinite horizontal plates induced by a pressure gradient or by relative motion of the plates in presence of an uniform magnetic field $\boldsymbol{B}_0$ imposed perpendicular to the plates. A Cartesian coordinate system is used in such a way that the z-axis is along the fluid flow direction and the x-axis is perpendicular to the plates so that the position of the plates are denoted by $x = +a$ and $x = -a$, respectively. The plates are assumed to be long enough so that the edge effects are negligible. Hence, all variables depend only on x, except the pressure p which may also vary in the direction of flow. Then we have the fluid velocity $\boldsymbol{V}(0, 0, v_z)$ and the magnetic field $\boldsymbol{B}(B_0, 0, B_z)$. Further we shall assume that the electric field is $\boldsymbol{E}(0, E_y, 0)$. Thus rot $\boldsymbol{E} = 0$ and Ohm's law yield

$$E_y = \text{Const.}\,, \qquad j_x = j_z = 0\,, \qquad j_y = \sigma(E_y + v_z B_0) \tag{1}$$

where σ is the electrical conductivity. Also we have $\partial p/\partial z = $ = Const. Taking into account these assumptions the z-component of the equation of motion together with (1) becomes

$$\mu \frac{d^2 v_z}{dx^2} - \sigma B_0^2 v_z = \left(\frac{\partial p}{\partial z} + \sigma E_y B_0\right) = -\sigma B_0^2 U \quad \text{(say)}, \tag{2}$$

in which μ is the dynamic viscosity of the fluid and U is a constant. The total current per unit length is

$$J_y = \frac{1}{2a}\int_{-a}^{+a} j_y \, dx = \sigma(E_y + B_0 v_m)$$

where

$$v_m = \frac{1}{2a}\int_{-a}^{+a} v_z \, dx\,.$$

If there is no external circuit then $J_y = 0$ and $E_y = -B_0 v_m$. Next, we shall consider the following problems:

(i) Poiseuille type flow. In this case the plates are stationary and therefore the boundary conditions for velocity are

$$v_z = 0 \quad \text{at} \quad x = \pm a\,. \tag{3}$$

The solution of Eq. (2) is then [2]

$$v_z = U\left(1 - \frac{\cosh \frac{Mx}{a}}{\cosh M}\right), \qquad v_m = U\left(1 - \frac{\tanh M}{M}\right) \tag{4}$$

where $M = aB_0(\sigma/\mu)^{1/2}$ is the Hartmann number.

Let us consider that the plates are kept at constant temperatures and the lower plate is maintained at a higher temperature, i.e.

$$T = T_0 \quad \text{at} \quad x = a \quad \text{and} \quad T = T_1 \quad \text{at} \quad x = -a \tag{5}$$

where $T_1 > T_0$. The energy equation becomes

$$\frac{d^2 T}{dx^2} = -\frac{\mu}{k}\left(\frac{dv_z}{dx}\right)^2 - \frac{\sigma B_0^2}{k}(v_z - v_m)^2 \tag{6}$$

including viscous dissipation and Ohmic dissipation. Here T is the temperature of the liquid and k is the thermal conductivity. The solution of (6) with boundary conditions (5) is

$$\Theta = \frac{1}{2}(1 - X) + \frac{Pr\,Ec}{2M(\cosh 2M + 1)}[M(X^2 - 1) \times \times (1 - \cosh 2M) + M(\cosh 2M - \cosh 2MX) + + 8 \sinh M(\cosh MX - \cosh M)] \tag{7}$$

where $\Theta = (T - T_0)/(T_1 - T_0)$, $X = x/a$, $Pr = \varrho \nu C_p/k$ is the Prandtl number and

$$Ec = \frac{U^2}{C_p(T_1 - T_0)} \tag{8}$$

is the Eckert number.

Using the results obtained above, we can easily calculate the net heat flux $q_w = -k(dT/dx)_{x=\pm a}$ from upper and lower plate to get

$$Nu|_{X=\pm 1} = \tfrac{1}{2} \pm Pr\,Ec \tanh M(M - \tanh M) \tag{9}$$

where the Nusselt number is defined as $Nu = q_w a/k(T_1 - T_0)$.

(ii) Couette type flow (when the plates are moving with equal and opposite velocities). The boundary conditions for the velocity are

$$v_z = \pm W \quad \text{at} \quad x = \pm a\,. \tag{10}$$

In this case we consider $U = 0$ and the solution of (2) subject to (10) is

$$v_z = W\frac{\sinh \frac{Mx}{a}}{\sinh M}\,.$$

Now, the energy equation (6) reduces to

$$\frac{d^2 T}{dx^2} = -\frac{\mu}{k}\left(\frac{dv_z}{dx}\right)^2 - \frac{\sigma B_0^2}{k} v_z^2\,. \tag{11}$$

The solution of (11) with the same boundary conditions (5) is

$$\Theta = \frac{1}{2}(1 - X) + \frac{Pr\,Ec}{2(\cosh 2M - 1)}(\cosh 2M - \cosh 2MX) \tag{12}$$

where now

$$Ec = \frac{W^2}{C_p(T_1 - T_0)}\,. \tag{13}$$

The heat transfer coefficient from upper and lower plates is

$$Nu|_{X=\pm 1} = \tfrac{1}{2} \pm Pr\,Ec\,M \operatorname{ctanh} M\,. \tag{14}$$

3. Discussion

The temperature profiles in the case of Poiseuille type flow, Eq. (7), are represented in Fig. 1 for different values of the Hartmann number M and $Pr\,Ec = 0.02$. It shows that the temperature of the liquid increases with the increase of the magnetic field M. This conclusion is true also for the case of Couette type flow. But, for the sake of saving space we do not give here this graphic.

The heat transfer coefficients (9) and (14) are displayed in Fig. 2 for upper and lower plates. Since for large values of M we have $\tanh M \to 1$ and $\operatorname{ctanh} M \to 1$, the heat transfer coefficients will be equal in both the case if the values of Prandtl

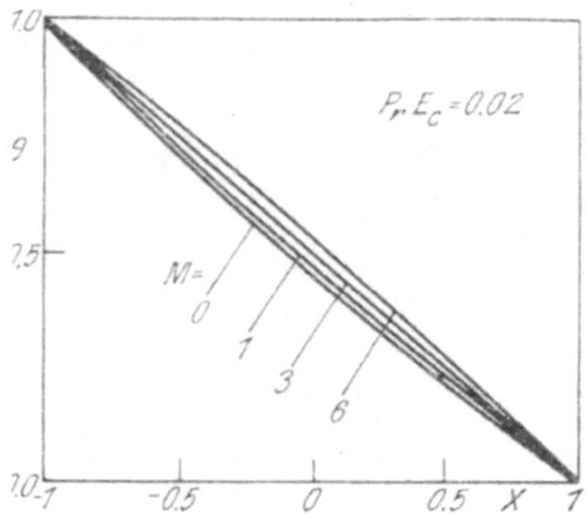

Fig. 1 Temperature profiles, Eq. (7)

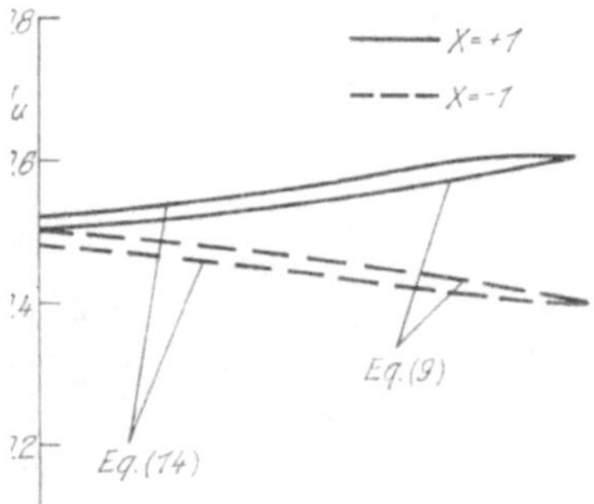

Fig. 2. Heat transfer coefficient for $Pr\,Ec = 0.02$

and Eckert numbers are equal. Hence, from (8) and (13) we see that $U = W$ as the Eckert numbers have equal values in both the cases. The velocity W of the plates in the Couette type flow must be equal to U, i.e.

$$-\frac{a^2}{\varrho\nu M \tanh M}\frac{\partial p}{\partial z}$$

in the Poiseuille type flow where the liquid motion is only due to pressure gradient. In other words, the pressure gradient in Poiseuille type flow must be maintained in such a way that

$$\frac{\partial p}{\partial z} = -\frac{\varrho\nu W}{a^2} M \tanh M\,.$$

An important conclusion is that for large values of M ($\gtrsim 5$) this relation gives

$$-\frac{a^2}{\varrho\nu W}\frac{\partial p}{\partial z} \to M\,.$$

References

1 Блум, Э. Я.; Михайлов, Ю. А.; Озолс, Р. Я., Тепло-и массообмен в магнитном поле, Рига „Зинатне„, 1980.

2 SHERCLIFF, J. A. A text book of magnetohydrodynamics, Pergamon Press, 1965.

Eingereicht: 23. März 1981

Anschriften. Dr. A. K. BORKAKATI, Department of Mathematics, Dibrugarh University, Dibrugarh-786004, India; Dr. I. POP, Faculty of Mathematics, University of Cluj, R-3400 Cluj, CP 253, Romania.

ZAMM 62, 52 – 54 (1982)

K. VAJRAVELU / L. DEBNATH

A Nonlinear Study of Combined Free and Forced Convection in Flows Due to a Rotating Disk

1. Introduction

VON KÁRMÁN [1] discovered that the steady state laminar flow of an incompressible viscous fluid due to the uniform rotation of an infinite disk inside an infinite reservoir can be described by an exact solution of the full Navier-Stokes equations. Using KÁRMÁN's solution of the Navier-Stokes equations, MILLSAPS and POHLHAUSEN [2] had shown that the decoupled but complete energy equation for the forced convection heat transfer for the same physical problem can be solved exactly.

In view of the widespread applicability of the viscous flow due to a rotating disk, LUK, MILLSAPS and POHLHAUSEN [3] showed an interesting aspect that the combined free and forced convection heat transfer problem can also be solved exactly when the equations of fluid dynamics and thermodynamics are coupled by operation of buoyancy with mercury as the working liquid. The fluid motion develops slowly following the rotation of the disk and the development of non uniformity in the temperature field. However, when the temperature differences are appreciably large, the volumetric heat generation or absorption term may exert strong influence on the heat transfer as a consequence, on the fluid flows as well.

The analysis of temperature field as modified by the generation or absorption of heat in moving fluids is important in view of several physical problems including (i) problems concerning chemical reaction (VERON [4]), (ii) problems dealing with dissociating fluids (LIGHTHILL [5] and LEES [6]). In fact, literature is replete with examples concerning the heat transfer in laminar flow of viscous fluids with internal heat generation or absorption.

In spite of the above works, an attention has hardly been given to the study of the effect of heat sources or sinks on the combined free and forced convection flows, and heat transfer problems such as convection studies with internal heat sources for Earth's mantle proposed by GASKELL [7], RUNCORN [8] and TOZER [9] and for the outer region of star interiors considered by BETHE [10].

In the present study, an attempt is made to throw some light on the effects of heat sources or sinks on the combined free and forced convection over a rotating disk with and without frictional heating effects, when the working fluid is either air or water (in practice, the most commonly used fluids). The primary objective of this paper is to solve the combined free and forced convection problem in fluid flows due to a rotating disk including the effects of internal heat generation. In order to incorporate all the above stated effects, we use the similarity transformations in conjunction with a nondimensionalization procedure different from that of LUK et al. This enables us to reduce the governing partial differential system to a nonlinear ordinary differential system, characterizing the combined free and forced convection over a rotating disk. The resulting nonlinear ordinary differential equations with relevant boundary conditions have been integrated numerically with great difficulty by the Runge-Kutta-Gill method with limited information available from the previous studies in giving initial values for the dependent flow and heat transfer variables. The numerical results thus obtained are presented in figures 1 through 5. Some discussion of the numerical results and a summary of the important conclusions are included.

2. Formulation and Solution of the Problem

The governing equations of the problem in cylindrical polar coordinates (r, θ, z) with corresponding velocity components $\{u(r, z), v(r, z), w(r, z)\}$ for fluid element at temperature $t(r, z)$ and the pressure $p(r, z)$ can be written as:

$$\frac{\partial u}{\partial r} + \frac{u}{r} + \frac{\partial w}{\partial z} = 0\,, \tag{1}$$

$$u\frac{\partial u}{\partial r} + w\frac{\partial u}{\partial z} - \frac{v^2}{r} = -\frac{1}{\varrho}\frac{\partial p}{\partial r} + \nu\left[\nabla^2 u - \frac{u}{r^2}\right], \tag{2}$$

$$u\frac{\partial v}{\partial r} + w\frac{\partial v}{\partial z} + \frac{uv}{r} = \nu\left[\nabla^2 v - \frac{v}{r}\right], \tag{3}$$

$$u\frac{\partial w}{\partial r} + w\frac{\partial w}{\partial z} = -\frac{1}{\varrho}\frac{\partial p}{\partial z} + \nu\nabla^2 w + g\beta\,(t - t_\infty), \tag{4}$$

$$u\frac{\partial t}{\partial r} + w\frac{\partial t}{\partial z} = \frac{k}{\varrho C_p}\nabla^2 t + \frac{\nu}{C_p}\varphi + \frac{Q^*}{\varrho C_p}(t - t_\infty), \tag{5}$$

where the Laplacian

$$\nabla^2 \equiv \frac{\partial^2}{\partial r^2} + \frac{1}{r}\frac{\partial}{\partial r} + \frac{\partial^2}{\partial z^2},$$

the dissipation function

$$\varphi \equiv 2\left(\frac{\partial u}{\partial r}\right)^2 + 2\left(\frac{u}{r}\right)^2 + 2\left(\frac{\partial w}{\partial z}\right)^2 + \left(\frac{\partial v}{\partial r} - \frac{v}{r}\right)^2 + \left(\frac{\partial v}{\partial z}\right)^2 + \left(\frac{\partial u}{\partial z} + \frac{\partial w}{\partial r}\right)^2,$$

and the last term in the energy equation is the heat source/sink term.

If ω denotes the angular velocity of the rotating disk, the boundary conditions of the problem are given by

$$u = 0, \quad v = r\omega, \quad w = 0, \quad t = t_d \quad \text{at} \quad z = 0, \tag{6}$$

$$u = 0, \quad v = 0, \quad t = t_\infty \quad \text{as} \quad z \to \infty. \tag{7}$$

Defining nondimensional variables

$$R = \left(\frac{\omega}{\nu}\right)^{1/2} r, \qquad Z = \left(\frac{\omega}{\nu}\right)^{1/2} z;$$

$$U = u/(\nu\omega)^{1/2}, \qquad V = v/(\nu\omega)^{1/2}, \qquad W = w/(\nu\omega)^{1/2};$$

$$P = p/\varrho\nu\omega, \qquad T = (t - t_\infty)/(t_d - t_\infty) \quad \text{and} \quad \Phi = \varphi/\omega^2;$$

equations (1)—(5) and the boundary conditions (6) and (7) can be written as:

$$\frac{\partial U}{\partial R} + \frac{U}{R} + \frac{\partial W}{\partial Z} = 0, \tag{8}$$

$$U\frac{\partial U}{\partial R} + W\frac{\partial U}{\partial Z} - \frac{V^2}{R} = -\frac{\partial P}{\partial R} + \nabla^2 U - \frac{U}{R^2}, \tag{9}$$

$$U\frac{\partial V}{\partial R} + W\frac{\partial V}{\partial Z} + \frac{UV}{R} = \nabla^2 V - \frac{V}{R^2}, \tag{10}$$

$$U\frac{\partial W}{\partial R} + W\frac{\partial W}{\partial Z} = -\frac{\partial P}{\partial Z} + \nabla^2 W + \gamma T, \tag{11}$$

$$U\frac{\partial T}{\partial R} + W\frac{\partial T}{\partial Z} = \frac{1}{\sigma}\nabla^2 T + E\Phi + \frac{1}{\sigma}\alpha T, \tag{12}$$

$$U = 0, \quad V = R, \quad W = 0, \quad T = 1 \quad \text{at} \quad Z = 0, \tag{13}$$

$$U = 0, \quad V = 0, \quad T = 0 \quad \text{as} \quad Z \to \infty \tag{14}$$

where $\gamma = (\nu/\omega)^{1/2}\, g\beta\Delta T/\nu\omega$, the free convection parameter,

$\sigma = \mu C_p/k$, the Prandtl number,

$E = \nu\omega/C_p\Delta T$, the Eckert number,

$\alpha = Q^*\nu/k\omega$, the heat source/sink parameter

and $\Delta T = t_d - t_\infty$, the temperature difference.

To reduce equations (8)—(12) to a system of ordinary differential equations, we use the substitutions similar to those of Luk, Millsaps and Pohlhausen [3] as follows:

$$U = RF(Z), \qquad V = RG(Z), \qquad W = H(Z); \tag{15}$$

$$P = R^2L(Z) + M(Z), \qquad T = R^2S(Z) + Q(Z). \qquad (16), (17)$$

In view of these transformations (15), (16) and (17), the system of five partial differential equations reduces into a set of seven ordinary differential equations:

$$2F + H' = 0, \qquad F^2 + HF' - G^2 = -2L + F'', \qquad (18), (19)$$

$$2FG + HG' = G'', \qquad HH' = -M' + H'' + \gamma Q, \qquad (20), (21)$$

$$L' = \gamma S, \qquad 2FS + HS' = \frac{1}{\sigma}S'' + E(F'^2 + G'^2) + \frac{1}{\sigma}\alpha S, \qquad (22), (23)$$

$$HQ' = \frac{1}{\sigma}(4S + Q'') + E(12F^2) + \frac{1}{\sigma}\alpha Q \tag{24}$$

where primes denote differentiation with respect to z. The boundary conditions then become

$$F = 0, \quad G = 1, \quad H = 0, \quad L = M = 0, \quad Q = 1,$$

$$S = 0 \quad \text{at} \quad Z = 0, \tag{25}$$

$$F = 0, \quad G = 0, \quad Q = 0, \quad S = 0 \quad \text{as} \quad Z \to \infty. \tag{26}$$

The above system of equations (18)—(24) with conditions (25) and (26) has been numerically integrated by the well-known Runge-Kutta-Gill method, with an integration range of $0 \leq Z \leq 10$. For experimental reasons, three values of 0.71 (air), 4(mercury), and 7 (water) are taken for the Prandtl number, and two values of the modified Grashof number $\gamma = 0$ and 0.01 are used. The velocity and the temperature fields have been calculated with $E = 0$, 0.02 and $\alpha = 0.05$, 0, -0.05 along with the above said values for γ and σ. The numerical results thus obtained are presented in figures 1 through 5.

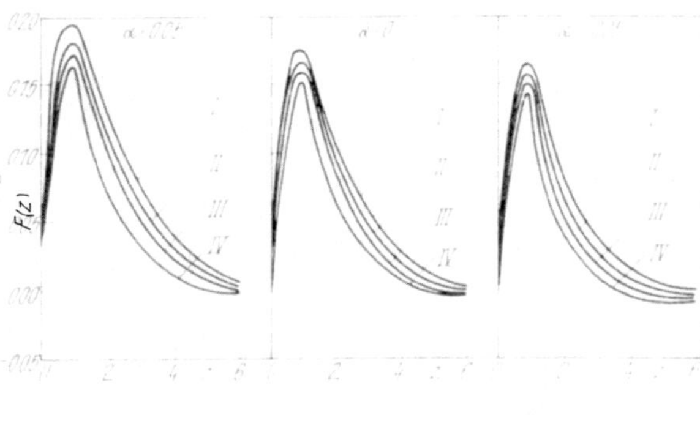

Fig. 1. Non-dimensional radial velocity component

	I	II	III	IV
γ	0.0	0.01	0.01	0.01
σ	0.71	0.71	4	7

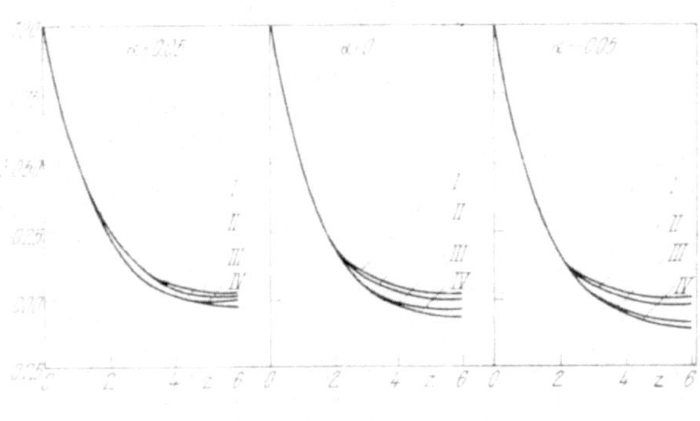

Fig. 2. Non-dimensional tangential velocity component. Curves I—IV as in figure 1

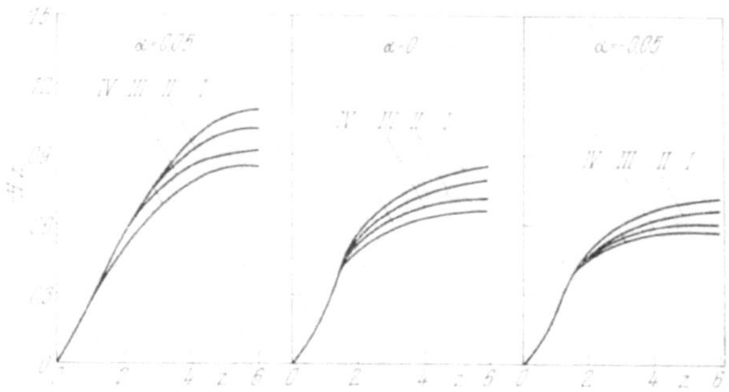

Fig. 3. Non-dimensional axial velocity component. Curves as in figure 1

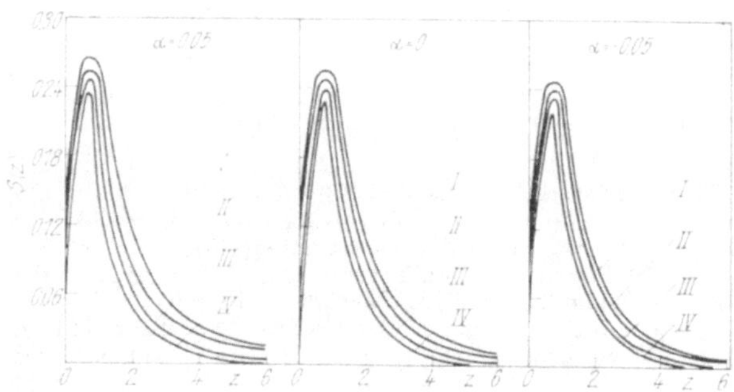

Fig. 4. Non-dimensional radial thermal distribution function. Curves I–IV as in figure 1

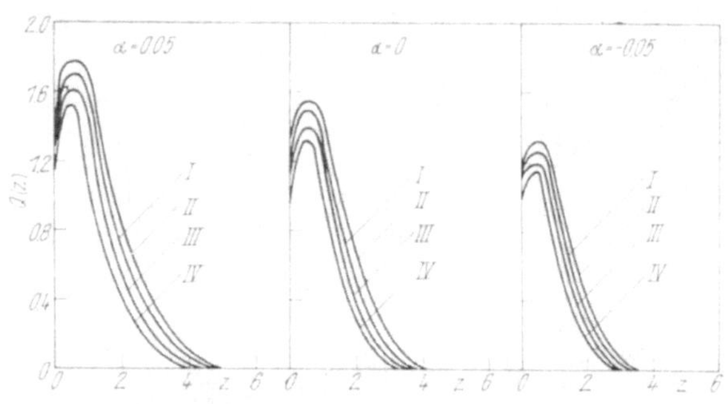

Fig. 5. Non-dimensional thermal distribution function Q vs. Z. Curves as in figure 1

3. Discussion of the Results

The equations (18)–(24) subject to the conditions (25) and (26) are solved by the Runge-Kutta-Gill method, mentioned in section 2. The results thus obtained are presented in figures 1–5 for $\sigma = 0.71$. 4, 7 and $\alpha = 0.05$, 0, -0.05 with $\gamma = 0$, 0.01 and $E = 0.02$. It is worth mentioning here that the results of Luk, Millsaps and Pohlhausen [3] for fluid velocity and fluid temperature are a special case of the solution presented in this paper (but for the different nondimensionalizations employed by Luck, Millsaps and Pohlhausen and the present authors) for $\sigma = 4$ and $\alpha = 0$. It is to be noted that the oscillatory nature of the solutions is not shown in figures; the amplitudes of the oscillations are insufficient in the computed cases to exhibit their oscillatory nature.

In figure 1 the numerical values for the radial velocity component F, for $\alpha = 0.05$, 0, -0.05, are shown with different Prandtl numbers. From figure 1 it can be observed that the values of velocity component F have increased considerably in the presence of heat sources. Opposite phenomenon is observed in the presence of heat sinks. Observations from figure 1 yielded that the velocity component F increases, as the Prandtl number decreases; the physical meaning of which is that the velocity component F is larger for the smaller Prandtl number fluids than the higher Prandtl number fluids. Also, the effect of the free convection parameter γ is found to decrease the radial velocity component F, both in the presence of heat sources and in their absence.

Figure 2 describes the behaviour of the tangential velocity component G. Here also, as in the case of F, the values of G increase with increasing α and decrease with increasing Prandtl number or the free convection parameter γ.

The behaviour of the axial velocity component H, perpendicular to the disk, is depicted in figure 3. The negative values of H, at all points in the fluid, indicate that the velocity H is directed towards the disk. Here also, as in the cases of F and G, the values of H (absolute) increase with increasing α and decrease with increasing Prandtl number or the free convection parameter γ.

The profiles for the thermal distribution functions S(radial), and Q(environmental and dissipative) are presented, respectively in figures 4 and 5. From these figures, one can easily observe that the values of S and Q are considerably increased in the presence of heat sources than in their absence. Reverse is the phenomenon in the presence of heat sinks. It can also be observed that the effect of Prandtl number σ is to decrease the values of S and Q, both in the presence of heat sources and in their absence. However, the free convection parameter γ increases the value of Q both in the presence of heat sources and in their absence. This result is duly reversed in the case of the radial temperature distribution S.

4. Conclusions

Some of the interesting conclusions of the problem are summarized as follows:

(i) the fluid velocity components F, G, H (absolute) and the temperature distributions S and Q increase considerably in the presence of heat sources than in their absence. This result is true even with the frictional heating parameter E;

(ii) in the absence of heat sources the velocity component H, perpendicular to the disk, is directed towards the disk. This observation is true even in the presence of heat sources or sinks;

(iii) the effect of the Prandtl number σ is found to decrease the fluid velocity components as well as the temperature distributions;

(iv) the effect of the free convection parameter γ is to decrease the values of the velocity components F, G, H (absolute), and the radial thermal distribution S. But the value of Q increases with γ;

(v) the oscillatory nature of the velocity components is observed, whenever buoyancy is taken into consideration.

References

1 Kármán, Th. von, Über laminare und turbulente Reibung, Z. angew. Math. Mech. **1**, 233 (1921).
2 Millsaps, K., and Pohlhausen, K., Heat transfer by laminar flow from a rotating plate, J. Aero. Sci. **19**, 120 (1952).
3 Luk, K. H., Millsaps, K., and Pohlhausen, K., An exact solution for the combined free and forced convection in flows due to rotating disks, Proc. 4th Int. Heat Transfer Conf., Versailles, Paper NC 4.2 (1970).
4 Veron, M., Bull. Tech. Soc. France, Const. Babcock and Wilcox. No. 21, Paris (1948).
5 Lighthill, M. J., F. M. 1958, Aero. Res. Council (1956).
6 Lees, L., Laminar heat over blunt-nosed bodies of hypersonic flight speeds, Jet Propulsion **26**, 259 (1956).
7 Gaskell, T. F. (ed.), The Earth's Mantle, New York Academic Press, (1967).
8 Runcorn, S. K., Convection currents in the Earth's mantle, Nature **195**, 1248 (1962).
9 Tozer, D. C., Heat transfer and convection current, Proc. Roy. Soc. (Lond.) **A 258**, 252 (1966).
10 Bethe, H. A., Energy Production in Stars, Science **16**, 541 (1968).

Eingereicht am 24. 12. 1980

Anschrift: K. Vajravelu / L. Debnath, Mathematics Department, East Carolina University, Greenville, North Carolina 27834, USA

ZAMM 62, 55 – 57 (1982)

C. PERDIKIS / G. TZIVANIDIS / A. RAPTIS

On Plane Deformations of Incompressible Isotropic Hyperelastic Material

1. Introduction

In 1978 [1] a formulation of the governing equations for finite plane-strain deformations of incompressible isotropic elastic solids was presented and two boundary-value problems for a circular annulus were examined. The main feature of that paper has been the use of an appropriate choice of deformation invariants different from those most commonly used [2].

Here we formulate the equations for finite plane-strain of incompressible isotropic hyperelastic solids, utilizing complex variables and employing the principal stretches rather than a particular choice of strain invariants. Several new results are obtained and some known results are recovered. All these are discussed in sections 2, 3, 4 and 5.

In section 6 the circular horizontal and uniform shear of a tube of such a material with OGDEN's strain-energy function is studied, not taking into account torsion. Some known results are also found. In section 7 we prove the uniqueness of the solution for this problem.

2. Kinematics of Deformation

In a fixed rectangular Cartesian coordinate system, the class of plane deformation fields with uniform transverse stretch is characterised in component form by

$$x_1 = x_1(X_1, X_2)\,, \qquad x_2 = x_2(X_1, X_2)\,, \qquad x_3 = a_3 X_3\,, \tag{1}$$

where X_i and x_i $(i = 1, 2, 3)$ are, respectively, the reference and deformed coordinates of a typical material point and a_3 is a positive constant. It is assumed that the undeformed configuration is free of stress. For such deformations the polar decomposition theorem $F = RU$, where R is a proper orthogonal matrix and U is a positive definite symmetric matrix, implies that the two dimensional deformation gradient matrix F can be written

$$\begin{pmatrix} \dfrac{\partial x_1}{\partial X_1} & \dfrac{\partial x_1}{\partial X_2} \\ \dfrac{\partial x_2}{\partial X_1} & \dfrac{\partial x_2}{\partial X_2} \end{pmatrix} = \frac{1}{2}(a_1 + a_2)\begin{pmatrix} \cos\omega & -\sin\omega \\ \sin\omega & \cos\omega \end{pmatrix} + \frac{1}{2}(a_1 - a_2)\begin{pmatrix} \cos\gamma & \sin\gamma \\ \sin\gamma & -\cos\gamma \end{pmatrix}. \tag{2}$$

In (2), the principal stretches a_1 and a_2 are the eigenvalues of the two dimensional strain matrix $(F^T F)^{1/2}$, where T denotes transposition; ω is the local angle of rotation of the material; $\Theta = \frac{1}{2}(\gamma - \omega)$ is the inclination to the X_1-axis of the local referential stretch axis in whose direction the principal stretch is a_1.

If we introduce the complex variables $z = x_1 + ix_2$, $Z = X_1 + iX_2$ and consider z as a function of Z and $\bar{Z}$ we find

$$\frac{\partial z}{\partial Z} = \frac{1}{2}(a_1 + a_2)\, e^{i\omega}, \qquad \frac{\partial z}{\partial \bar{Z}} = \frac{1}{2}(a_1 - a_2)\, e^{i\gamma}. \tag{3}$$

These results are analogous to those of FOSDICK and SCHULER [3, section 5].

3. Analysis of Stress

The in-plane components of the first Piola-Kirchhoff stress tensor (the transpose of the nominal stress tensor $\mathbf{S}$) form the matrix [4]

$$S^T = \begin{pmatrix} S_{11} & S_{21} \\ S_{12} & S_{22} \end{pmatrix} = RUt\,, \tag{4}$$

where t is the second Piola-Kirchhoff stress matrix and where use has been made of the polar decomposition theorem. It follows that S^T has a representation directly analogous to (2)

$$S^T = \begin{pmatrix} S_{11} & S_{21} \\ S_{12} & S_{22} \end{pmatrix} = \frac{1}{2}(S_1 + S_2)\begin{pmatrix} \cos\omega & -\sin\omega \\ \sin\omega & \cos\omega \end{pmatrix} + \frac{1}{2}(S_1 - S_2)\begin{pmatrix} \cos\gamma & \sin\gamma \\ \sin\gamma & -\cos\gamma \end{pmatrix}, \tag{5}$$

where, in general, S_1 and S_2 are not the principal values of S.

In the absence of body forces the equilibrium equations, in referential form, are

$$S_{11,1} + S_{21,2} = 0\,, \qquad S_{12,1} + S_{22,2} = 0\,. \tag{6}$$

Equations (6) imply the existence of a complex stress function $\Phi = \Phi_1 + i\Phi_2$, whose components Φ_1 and Φ_2 generate the components $S_{\alpha\beta}$ $(\alpha, \beta = 1, 2)$ through the relations

$$S_{11} = -\Phi_{1,2}\,, \quad S_{12} = -\Phi_{2,2}\,, \quad S_{21} = \Phi_{1,1}\,, \quad S_{22} = \Phi_{2,1}\,. \tag{7}$$

If we consider Φ as a function of Z and $\bar{Z}$ we obtain

$$\frac{\partial \Phi}{\partial Z} = \frac{1}{2} i(S_1 + S_2)\, e^{i\omega}, \qquad \frac{\partial \Phi}{\partial \bar{Z}} = -\frac{1}{2} i(S_1 - S_2)\, e^{i\gamma}. \tag{8}$$

4. Constitutive Equations

If the isotropic incompressible hyperelastic material under consideration possesses a strain-energy function $W(a_1, a_2)$, per unit volume in the undeformed configuration, the constitutive equations take the form

$$S_1 = -pa_1^{-1} + \frac{\partial W}{\partial a_1}, \qquad S_2 = -pa_2^{-1} + \frac{\partial W}{\partial a_2}, \tag{9}$$

where the pressure p is not determined a priori by the deformation and it arises as a reaction to material incompressibility. Since $a_1 a_2 = a_3^{-1}$, because of the incompressibility, W is expressible as a function of a single strain invariant (see GREEN and ADKINS [2])

$$\begin{aligned} W(\mathrm{I}, \mathrm{II}, \mathrm{III}) &= W(a_1^2 + a_2^2 + a_3^2,\, a_2^2a_3^2 + a_3^2a_1^2 + a_1^2a_2^2,\, 1) = \\ &= W(a_1^2 + a_2^2 + a_3^2,\, a_3^2(a_1^2 + a_2^2) + a_3^{-2}) = W(a_1, a_2) = \\ &= F(a_1^2 + a_2^2)\,, \end{aligned} \tag{10}$$

where I, II, III are the principal invariants of both Cauchy-Green strain tensors $\mathbf{C}$ and $\mathbf{B}$ [5]. Then, equations (9) are written

$$S_1 = -pa_1^{-1} + 2a_1 F'\,, \qquad S_2 = -pa_2^{-1} + 2a_2 F'\,, \tag{11}$$

where a dash denotes differentiation with respect to the invariant $I_1 = a_1^2 + a_2^2$.

5. Governing Equations

The elimination of the angles ω and γ between (3) and (8) yields

$$\frac{\partial \Phi}{\partial Z} = i\,\frac{S_1 + S_2}{a_1 + a_2}\,\frac{\partial z}{\partial Z}, \qquad \frac{\partial \Phi}{\partial \bar{Z}} = -i\,\frac{S_1 - S_2}{a_1 - a_2}\,\frac{\partial z}{\partial \bar{Z}}, \tag{12}$$

and from these we derive the partial differential equations

$$\begin{aligned} &\frac{\partial}{\partial Z}\left(\frac{S_1 - S_2}{a_1 - a_2}\,\frac{\partial z}{\partial \bar{Z}}\right) + \frac{\partial}{\partial \bar{Z}}\left(\frac{S_1 + S_2}{a_1 + a_2}\,\frac{\partial z}{\partial Z}\right) = 0\,, \\ &\frac{\partial}{\partial Z}\left(\frac{a_1 - a_2}{S_1 - S_2}\,\frac{\partial \Phi}{\partial \bar{Z}}\right) + \frac{\partial}{\partial \bar{Z}}\left(\frac{a_1 + a_2}{S_1 + S_2}\,\frac{\partial \Phi}{\partial Z}\right) = 0\,, \end{aligned} \tag{13}$$

for z and Φ respectively.

From $(13)_1$ and (11) we deduce

$$\frac{\partial}{\partial Z}\left(p_1\frac{\partial z}{\partial \bar{Z}}\right) + \frac{\partial}{\partial \bar{Z}}\left\{(2W_1 - p_1)\frac{\partial z}{\partial Z}\right\} = 0 \tag{14}$$

where

$$p_1 = 2F' + a_3 p = W_1 + a_3 p\,,$$

and it is equation (14) from which we work in this paper.

6. Circular Shear of a Tube of Incompressible Isotropic Hyperelastic Material with OGDEN's Strain-Energy Function

We consider a circular cylindrical tube of radii A and B $(B > A)$ and height L in the undeformed configuration. We assume that the inner surface $R = A$ is held fixed, while a shear traction $t_\theta = S$ is applied horizontally and uniformly to the external surface $R = B$. Let $r = b$ be the radius of the external surface and l the height of the tube in the deformed configuration.

Since the material is assumed incompressible, it follows that

$$r^2 = \frac{R^2 - (1 - a_3) A^2}{a_3}, \tag{15}$$

where R and r are the radii of any material point of the body in the undeformed and deformed configuration respectively, and $a_3 = \frac{l}{L}$. If the plane polar coordinates of any material point, before and after deformation, are (R, Θ) and (r, θ) respectively, then

$$Z = \mathrm{R}e^{i\Theta}, \qquad z = \mathrm{r}\, e^{i\theta}. \tag{16}$$

Since z is a function of Z and $\overline{Z}$, and r is a function of R, it follows that $\theta = \theta(R, \Theta)$.

The incompressibility condition $J = \left|\frac{\partial x_i}{\partial X_j}\right| = 1$ $(i, j = 1, 2, 3)$ can be written in the form

$$\frac{\partial\theta}{\partial\Theta} = 1, \tag{17}$$

using the relations $x_1 = r\cos\theta$, $x_2 = r\sin\theta$ and (15). If we do not take into account torsion, the solution of this equation is

$$\theta = \Theta + \varphi(R), \tag{18}$$

where φ is an arbitrary function of R finally determined from the boundary conditions. Evidently, a boundary condition is $\varphi(A) = 0$. With the help of (18) we obtain

$$z = r\frac{Z}{R} e^{i\varphi}, \tag{19}$$

and the substitution of the partial derivatives $\frac{\partial z}{\partial Z}$ and $\frac{\partial z}{\partial \overline{Z}}$ into (14) leads to the relations

$$(r^2 R W_1 \varphi')' = 0, \qquad \varphi'' + \left(\frac{W_1'}{W_1} + \frac{2R}{a_3 r^2} + \frac{1}{R}\right)\varphi' = 0, \tag{20}$$

where a prime denotes differentiation with respect to R, $W_1' = \frac{\mathrm{d}W_1}{\mathrm{d}R} = \frac{\mathrm{d}}{\mathrm{d}R}(2F') = \frac{\mathrm{d}}{\mathrm{d}R}\left(\frac{\mathrm{d}W}{\mathrm{d}I_2}\right)$ and $I_2 = \frac{1}{2}\, I_1$ is a deformation invariant [1]. Equation $(20)_1$ is reduced to the equation (53) given by OGDEN [1] when $a_3 = 1$.

We can express the strain-energy function W in terms of a variable quantity, say a. If we put $a_1 = a_3^{-1/2}a$, $a_2 = a_3^{-1/2}a^{-1}$, where $a = a(R)$, then the incompressibility condition is satisfied. The invariant I_2 becomes·

$$I_2 = \frac{1}{2} a_3^{-1}(a^2 + a^{-2}), \tag{21}$$

and thus W is expressed in the form $W = W(a)$. Hence we have

$$W_1 = \frac{a_3 a}{a^2 - a^{-2}}\frac{\mathrm{d}W}{\mathrm{d}a}. \tag{22}$$

Equation $(20)_1$ is now valid with W_1 given by (22)

$$r^2 R \frac{a_3 a}{a^2 - a^{-2}}\frac{\mathrm{d}W}{\mathrm{d}a}\varphi' = C, \tag{23}$$

where C is an arbitrary constant. Equation $(20)_2$ determines the pressure after the deformation has been found.

We consider an incompressible isotropic hyperelastic solid for which there is a strain-energy function of the type proposed by OGDEN [6] to describe the large elastic deformations of rubber-like solids under isothermal conditions. The strain-energy function has the form

$$W = \sum_{n=1}^{s} \frac{\mu_n}{\alpha_n}(a_1^{\alpha_n} a + {}_2^{\alpha_n} + a_3^{\alpha_n} - 3), \tag{24}$$

where the α_n and μ_n are material constants with α_n dimensionless and μ_n with dimensions of stress. This equation is written

$$W = \sum_{n=1}^{s} \frac{\mu_n}{\alpha_n}\{a_3^{-\alpha_n/2}(a^{\alpha_n} + a^{-\alpha_n}) + a_3^{\alpha_n} - 3\}, \tag{25}$$

and (23) yields

$$a_3 r^2 R\varphi' \sum_{n=1}^{s} \mu_n a_3^{-\alpha_n/2}\frac{a^{\alpha_n} - a^{-\alpha_n}}{a^2 - a^{-2}} = C. \tag{26}$$

The two principal stretches, except of a_3, are the positive roots of the determinant

$$\begin{vmatrix} f^2 - \gamma^2 & fg \\ fg & g^2 + h^2 - \gamma^2 \end{vmatrix} - \gamma^4 - (f^2 + g^2 + h^2)\gamma^2 + h^2 f^2 = 0, \tag{27}$$

where, in general, $f = \frac{\mathrm{d}r}{\mathrm{d}R}$, $g = r\frac{\partial\theta}{\partial R}$ and $h = \frac{r}{R}\frac{\partial\theta}{\partial\Theta}$. We can show it, by observing that the deformation gradient matrix can be written in the form

$$F_{\alpha\beta} = \begin{pmatrix}\cos\theta & -\sin\theta \\ \sin\theta & \cos\theta\end{pmatrix}\begin{pmatrix} f & 0 \\ g & h\end{pmatrix}\begin{pmatrix}\cos\Theta & \sin\Theta \\ -\sin\Theta & \cos\Theta\end{pmatrix}$$

which in tensor form is

$$\boldsymbol{F} = \boldsymbol{P}^T\hat{\boldsymbol{F}}\boldsymbol{P}_0.$$

The left Cauchy-Green strain tensor $\boldsymbol{B}$ becomes, $\boldsymbol{B} = \boldsymbol{P}^T\hat{\boldsymbol{B}}\boldsymbol{P}$, where $\hat{\boldsymbol{B}} = \hat{\boldsymbol{F}}\hat{\boldsymbol{F}}^T$ and thus $\boldsymbol{B}$ and $\hat{\boldsymbol{B}}$ have the same principal stretches.

If the positive roots of (27) are γ_1 and γ_2, and we recall that the principal stretches are given by $a_1 = a_3^{-1/2}a$, $a_2 = a_3^{-1/2}a^{-1}$, we get

$$\gamma_1^2 + \gamma_2^2 = a_3^{-1}(a^2 + a^{-2}) = f^2 + g^2 + h^2.$$

We can express the quantity $a^2 + a^{-2}$ in terms of R, through the relation

$$a_3^{1/2} r\varphi' = \left\{(a^2 + a^{-2}) - \left(\frac{R^2}{a_3 r^2} + \frac{a_3 r^2}{R^2}\right)\right\}^{1/2}, \tag{28}$$

using that $f = \frac{R}{a_3 r}$, $g = r\varphi'(R)$ and $h = \frac{r}{R}$.

Equations (28) and (26) determine φ' and a, and φ follows from

$$\varphi(R) = \int_A^R \varphi'(R)\,\mathrm{d}R. \tag{29}$$

7. Uniqueness of the Solution

We go over the analysis of deformation in order to see if the relation (28) between a, R and φ' can be simplified. We distinguish two cases

1) Let $a_3 = 1$. Then $r = R$ and (28) yields

$$r\varphi' = a - a^{-1}. \tag{30}$$

This is the simplified relation between a, R and φ'. Equation (26) becomes

$$\sum_{n=1}^{s} \mu_n \frac{a^{\alpha_n} - a^{-\alpha_n}}{a + a^{-1}} = \frac{C}{R^2}. \tag{31}$$

This is a relation between a and R. If $\mu_n\alpha_n > 0$ (for each n) and $|\alpha_n| > 1$ (for each n), then the left-hand side of (31) is a strictly increasing function of a $\forall a > 1$ and therefore with $C > 0$ equation (31) must determine a unique positive $a(R)$ and hence, via (30), a unique $\varphi(R)$ [7].

2) Let $a_3 \neq 1$. Then equation (15) is written

$$a_3\frac{r^2}{R^2} = 1 + \beta, \tag{32}$$

where

$$\beta = (a_3 - 1)A^2/R^2. \tag{33}$$

The simplified relation is

$$a_3 r^2 \varphi'^2 = (a - a^{-1}) - \frac{\beta^2}{1+\beta}, \tag{34}$$

and (26) becomes

$$\sum_{n=1}^{s} \mu_n a_3^{-\alpha_n/2} \frac{a^{\alpha_n} - a^{-\alpha_n}}{a + a^{-1}} = \frac{\beta C}{(a_3 - 1) A^2} \times$$

$$\times \left\{1 + \beta - \frac{\beta^2}{(a - a^{-1})^2}\right\}^{-1/2} \tag{35}$$

This is a relation between a and β (or R). We examine if (35) has a unique solution for a given R. We can easily see that the left-hand side, say $f(a)$, is a strictly increasing function of a. We denote the right-hand side as $g(a)$. Let

$$\alpha = \frac{a_3^{1/2} r}{R} = \left\{1 + (a_3 - 1) \frac{A^2}{R^2}\right\}^{1/2}. \tag{36}$$

If

$$1)\ a_3 > 1 \text{ then } \alpha > 1, \quad 2)\ a_3 = 1 \text{ then } \alpha = 1,$$

$$3)\ 0 < a_3 < 1 \text{ then } 0 < \alpha < 1. \tag{37}$$

The function $g(a)$ is written

$$g(a) = \frac{K\left(1 - \frac{1}{a^2}\right)}{\alpha\left(1 - \frac{\alpha^2}{a^2}\right)^{1/2}\left(1 - \frac{\alpha^{-2}}{a^2}\right)^{1/2}}, \tag{38}$$

where

$$K = CR^{-2}. \tag{39}$$

We define

$$\hat{a} = \left(\frac{a_3^{1/2} r}{R}\right)^{\operatorname{sgn}(a_3 - 1)} \begin{cases} \alpha & \text{if } a_3 > 1, \\ \alpha^{-1} & \text{if } 0 < a_3 < 1, \end{cases} \tag{40}$$

where sgn denotes the signum function. Then $g(a)$ is defined for $a > \hat{a}$ (> 1) and $g(a) \to \infty$ as $a \downarrow \hat{a}$. Also, $g(a) \to K\alpha^{-1}$ as $a \to \infty$ and $g'(a) < 0\ \forall\ a > \hat{a}$. The graph of $g(a)$ is thus as shown

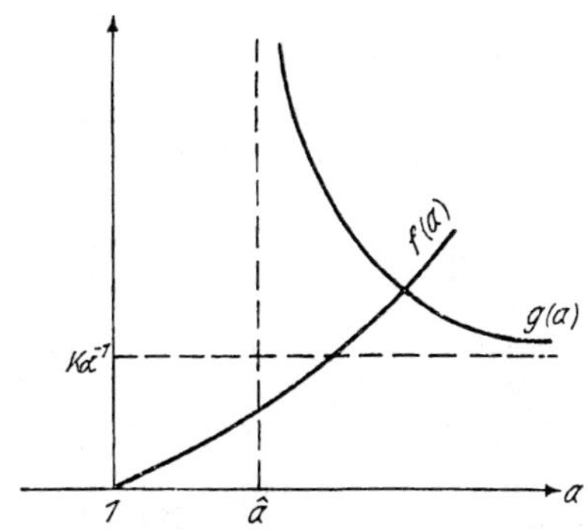

Fig. 1. Variation of the functions $f(a)$ and $g(a)$

(Fig. 1) and it clearly intersects the graph of $f(a)$ in precisely one point. The value of a at the point of intersection exceeds $\hat{a}$.

Therefore, equation (35) must determine a unique $a(R)$ and hence, via (34), a unique $\varphi(R)$.

References

1 OGDEN, R. W., On plane deformations of incompressible isotropic elastic solids, Math. Proc. Camb. Phil. Soc. **83**, pp. 127–136 (1978).
2 GREEN, A. E.; ADKINS, J. E., Large Elastic Deformations, Clarendon Press, Oxford (1970).
3 FOSDICK, R. L.; SCHULER, K. W., On ERICKSEN's problem for plane deformations with uniform transverse stretch, Int. J. Engng. Sci. **7**, pp. 217–233 (1969).
4 CHADWICK, P., Continuum Mechanics, George Allen and Unwin Ltd, London (1976).
5 TRUESDELL, C.; NOLL, W., The Non-Linear Field Theories of Mechanics, FLÜGGE's Handbuch der Physik, III/3, Springer (1965).
6 OGDEN, R. W., Large deformation isotropic elasticity — on the correlation of theory and experiment for incompressible rubber-like solids, Proc. R. Soc. Lond. A. **326**, pp. 565–584 (1972).
7 OGDEN, R. W.; CHADWICK, P., On the deformation of solid and tubular cylinders of incompressible isotropic elastic material, J. Mech. Phys. Solids **20**, pp. 77–90 (1972).

Eingereicht am 17. 12. 1980

Anschrift: Dr. C. PERDIKIS, Prof. G. TZIVANIDIS and Dr. A. RAPTIS University of Ioannina, School of Physics and Mathematics, Ioannina, Greece

BUCHBESPRECHUNGEN

Fischer, U. / Stephan, W., Mechanische Schwingungen. Leipzig, VEB Fachbuchverlag 1980. 332 S., M 22,–. BN 5465338

In der vorgelegten Monographie wird das Gesamtgebiet der Schwingungsmechanik für Studierende des Maschinenbaues und für in der Praxis auf diesem Gebiet tätige Ingenieure in moderner Form lehrbuchmäßig dargestellt. Es ist didaktisch klar, anschaulich und für den angesprochenen Leserkreis mathematisch angemessen formuliert. Allerdings beschränkt es sich auf rein mechanische Schwingungsprobleme und spart z. B. elektromechanische Schwingungssysteme völlig aus. Nach einer Einführung in die mathematischen Beschreibungsmöglichkeiten von Schwingungen werden einfache lineare Schwinger und die dahin führenden Modellbildungen ausführlich behandelt. Dabei wird eine methodisch und phänomenologisch umfassende Darstellung geboten. So sind z. B. nicht-stationäre Vorgänge, die Laplace-Transformation und stochastische Schwingungen darin enthalten, und auch Aufgaben werden bis zum zahlenmäßigen Endergebnis durchgerechnet. Es schließt sich dann ein umfangreicher Abschnitt über nicht-lineare Schwingungen mit einem Freiheitsgrad an. In ihm wird ein Überblick über die Lösungsmethoden gegeben, der durch die Vielzahl der aufgenommenen Methoden besonders bemerkenswert ist. So findet man z. B. neben der Energie-Methode und der Methode der Anstückelung die äquivalente Linearisierung, die Störungsrechnung, das Verfahren von GALERKIN und das Verfahren von BOGOLJUBOV und MITROPOLSKIJ. In dem Kapitel über lineare Systeme mit mehreren Freiheitsgraden wird die Aufstellung der Schwingungsgleichungen über die Lagrangeschen Gleichungen, mit der Kraftgrößenmethode und mit der Deformationsmethode vorgeführt. Im weiteren steht dann die Matrizeneigenwertaufgabe im Vordergrund, und man findet über den Rayleighschen Quotienten hinaus das Verfahren von JACOBI, die Vektoriteration nach V. MISES und auch Abschätzungen der Eigenfrequenzen. Für die nichtlinearen Systeme ist die Systematik schwieriger, doch versucht das Buch erfolgreich den Anschluß an die für einen Freiheitsgrad dargestellten Methoden. Für die linearen parametererregten Systeme wird in einem Abschnitt immerhin die Floquetsche Theorie vorgestellt. Die abschließenden Kapitel über die Schwingungen der Kontinua und über die Stabilität können die Fragestellungen dieser umfangreichen Gebiete naturgemäß nur andeuten.

Die vorliegende Darstellung der mechanischen Schwingungen für Studierende und Ingenieure des Maschinenbaues ist besonders von der Vielzahl der vorgestellten Fragestellungen und Methoden her interessant und aktuell. Sie müssen verständlicherweise bei der Breite ihrer Zielsetzung gelegentlich auf detaillierte Herleitungen verzichten, doch findet das Buch hierbei durchaus eine für die Leserzielgruppe nach Inhalt und Umfang angemessene Form.

Karlsruhe F. WEIDENHAMMER

Klotter, K., Technische Schwingungslehre. Band 1: Einfache Schwinger. 3., völlig neubearb. und erw. Auflage. Herausgegeben mit Unterstützung durch G. Benz. Teil B: Nichtlineare Schwingungen. Berlin-Heidelberg-New York, Springer-Verlag 1980. XVI, 576 S., 211 Abb., DM 98,–. US $ 57.90. ISBN 3-540-09327-3

Der nun vorliegende Teil B über nichtlineare Schwingungen mit einem Freiheitsgrad wurde von vielen, die die ersten beiden Auflagen aus den Jahren 1938 und 1951 als ein oder das Standardwerk der Schwingungstechnik benutzt haben, mit Spannung erwartet: Wird es angesichts der Breite und Vielschichtigkeit der Forschung, die zu zahlreichen Konferenzen und Monografien aus diesem Gebiet geführt hat, gelingen, an die besondere Bedeutung der früheren Auflagen anzuknüpfen?

In der offenbaren Erkenntnis, daß ein enzelnes Buch heutzutage nur Teilaspekte dieses komplexen Forschungsgebietes behandeln kann, wurde auf eine enzyklopädische Ergänzung der früheren Darstellungen verzichtet und ein gut ausgewählter Kreis von Methoden und Ergebnissen klar und in verständlicher Breite dargestellt. So ist ein völlig neues Buch entstanden, neben dem man auch weiterhin die Erstauflagen wird benutzen können und das andererseits die besser lesbare Form des normalen Buchdruckes verdient gehabt hätte.

Über die Hälfte des Buches ist autonomen Schwingungen gewidmet, dabei ist die Darstellung so angelegt, daß sie weitgehend auf nichtautonome Schwingungen übertragbar ist. Ausführlich mit Beispielen behandelt wird die Einführung dimensionsloser Größen. Eingehend dargestellt werden Phasenkurve und Grenzzykel, besonders genannt seien Schwinger mit Schaltern und Näherungsverfahren zur Bestimmung der Phasenkurven und der Differentialgleichungen im Zeitbereich. Zwischen primärer Näherung und vollständiger erster Näherung wird sorgfältig unterschieden. Es wird gezeigt, daß das Mittelungsverfahren einfacher anwendbar ist als das Störungsverfahren, aber offen läßt, ob die „erste" Näherung eine echte Näherung erster Ordnung ist. Im Rahmen der primären Näherung ergibt sich, daß Dämpfungskräfte, die der n-ten Potenz der Auslenkung, der Geschwindigkeit oder der Amplitude proportional sind, gleichwertig sind, so daß viele der Diskussionen über den „richtigen" Dämpfungsansatz ohne Bedeutung sind.

Im zweiten, nichtautonome Schwingungen behandelnden Teil des Buches wird zunächst an der Bezeichnung „Resonanzkurve" Kritik geübt und dafür der aus dem Englischen übernommene Ausdruck „Responsekurve" (Amplituden-Responsekurve und analog dazu Nullphasenwinkel-Responsekurve) eingeführt. Für eine Vielzahl verschiedener nichtlinearer Rückstell- und Dämpfungsfunktionen werden die Responsekurven berechnet und diskutiert. Stark nichtlineare nichtautonome Differentialgleichungen werden einmal in der Form pseudo-autonomer (abschnittsweise autonomer) Systeme im Zeitbereich und in der Phasenebene, zum anderen in der Form stückweise linearer Systeme behandelt, wobei ein auf einer schwingenden Platte hüpfender Ball, Stoßschwingungsdämpfer und Schwinger mit Reibkräften betrachtet werden. Der letzte Abschnitt über aktive (selbst- und fremderregte) Systeme enthält zahlreiche neue Ergebnisse über die harmonische, sub- und superharmonische Mitnahme.

Das Buch ist Lernenden ebenso wie in der Forschuung oder Praxis Tätigen als ein grundlegendes Werk auf dem Gebiet der nichtlinearen Schwingungen sehr zu empfehlen.

Berlin G. Schmidt

Müller, P. C. / Schiehlen, W. O., Forced Linear Vibrations. CISM Courses and Lectures 172. Wien-New-York, Springer-Verlag 1977. 146 S., 35 Abb., DM 36,–. S 248,– US S 19.80. ISBN 3-211-81487-6

Ein Lehrbuch über das relativ enge Gebiet linearer erzwungener Schwingungen ist sicherlich ungewöhnlich; man kann deshalb wohl Besonderes erwarten und wird darin auch nicht enttäuscht. Das Buch gibt im wesentlichen den Inhalt der Vorträge wieder, die beide Verfasser im Jahre 1973 am International Centre for Mechanical Sciences (CISM) in Udine (Italien) gehalten haben. Es beginnt mit einer kurzen Zusammenfassung der Methoden für die mathematische Formulierung der Differentialgleichungen und der allgemeinen Lösung. Es folgt eine Darstellung verschiedener Stabilitätsdefinitionen und ihre Anwendung auf lineare Schwingungssysteme. Dem schließt sich die Untersuchung deterministisch erregter Schwingungen an, wobei verständlicherweise die harmonisch erregten Schwingungen den größten Raum einnehmen. Insbesondere werden die Erscheinungen der Resonanz, Scheinresonanz und Tilgung untersucht. Den Abschluß des Hauptteiles bilden Ausführungen zur Korrelationstheorie stochastisch erregter Schwingungen. Für den Fachmann interessant ist auch der Anhang, der einige neuere Betrachtungen zur Steuerbarkeit und Beobachtbarkeit linearer Systeme sowie zur Existenz der Ljapunov-Funktion enthält. Der Inhalt des Buches ist klar gegliedert, die mathematische Formulierung ist exakt und leicht verständlich. Nach Meinung des Rezensenten bedaf es dennoch gründlicher Kenntnisse der Technischen Mechanik und auch der Anfangsgebiete der Schwingungslehre, um sich das Dargelegte voll zu eigen machen und es auch anwenden zu können. Insbesondere sind die Fragen der Modellfindung und Modellbeschreibung sehr knapp behandelt. Angaben oder Anleitungen zur Abschätzung der Modelleigenschaften, der Bewältigung der Datenmengen und überhaupt numerische Methoden fehlen ganz. Für den theoretisch Arbeitenden stellt das Buch jedoch eine wertvolle Ergänzung der bestehenden Literatur dar, der er manche neue Anregung verdanken wird.

Magdeburg U. Fischer

Wesolowski, Z. (Ed.), Nonlinear Dynamics of Elastic Bodies. CISM Courses and Lectures 227. Wien-New York, Springer-Verlag 1978. V, 212 S., 38 Abb., DM 39,– S. 277,–. ISBN 3-211-81512-0

In dem Band werden 5 Beiträge zu unterschiedlichen Problemkreisen zusammengefaßt, die der nichtlinearen Elastodynamik unterzuordnen sind. Bei der systematischen Zusammenstellung der benötigten Grundbegriffe zur Elastizität und Thermoelastizität wird im Übersichtsartikel von T. Manacorda „Topics in Elastodynamics" auf einige Tendenzen der neueren Forschung eingegangen. So werden interessante Beispiele für unterschiedliche Regularitätsbegriffe angegeben, das Signorini-Problem im nichtlinearen dynamischen Fall formuliert sowie die Müllersche Verallgemeinerung des zweiten Hauptsatzes der Thermodynamik betrachtet. Weitere Abschnitte befassen sich z. B. mit dem elastischen Potential und der Linearisierung nach Signorini. In diesem Zusammenhang finden Resultate von Capriz und P. Guidugli über Bedingungen zur Kompatibilität des nichtlinearen Problems mit den bei der Linearisierung zu untersuchenden linearen Hilfsproblemen Erwähnung. In C. Wozniaks Beitrag „Analytical Mechanics of Elastic Media" wird das Kontinuum als Gesamtheit materieller Partikel mit gewissen constraints aufgefaßt. Einige Spezialfälle solcher constraints werden betrachtet, die auf bekannte Theorien führen. Im Rahmen dieses Zugangs werden die Prinzipien der virtuellen und komplementären virtuellen Arbeit sowie die Lagrangeschen Gleichungen erster und zweiter Art betrachtet. Z. Wesolowski analysiert in „Wave Propagation in Finitely Deformed Elastic Material" verschiedene mögliche Formen der Wellenausbreitung im Rahmen der nichtlinearen Theorie. Insbesondere werden z. B. Beschleunigungswellen, deren Reflektion und Brechung, Schallwellen, sphärische Wellen in isotropen Materialien und Geschwindigkeitswellen untersucht. In der Darstellung wird konsequent der Tensorkalkül verwendet. Die Arbeit wird durch viele Diagramme bereichert, wie auch der folgende sehr ansprechende Beitrag „On the Non-Linear Behaviour and the Stability of Reticulated Elastic Systems" von S. J. Britvec. Den Ausgangspunkt für die Untersuchungen an Stabtragwerken stellt hier die totale potentielle Energie des Systems dar. Hieraus werden die benötigten Informationen über die Gleichgewichtsbahnen sowie das Stabilitätsverhalten in Abhängigkeit von einem Lastparameter insbesondere in der Nähe der einfachen Verzweigungspunkte gewonnen. Neben perfekten elastischen Systemen (d. h., das System ist vollständig durch die verallgemeinerten Koordinaten der Deformation bestimmt) werden auch Probleme mit imperfekten Bedingungen studiert. Im letzten thematisch etwas abweichenden Artikel „Generalized Measures for Large Deformations" von B. R. Seth wird eine Übersicht gegeben über Theorie und Anwendung verallgemeinerter Deformationsmaße in verschiedenen Zweigen der Kontinuumsmechanik, wobei besonders ausführlich auf Kriechdeformationen eingegangen wird. Alle Beiträge werden ergänzt durch Literaturverzeichnisse, auf die der Leser oft verwiesen wird, wenn bei der notwendig knapp gehaltenen Darstellungsweise ausführliche Betrachtungen im Text nicht möglich sind.

Leipzig J. Maul

Naudascher, E. / Rockwell, D. (eds.), Practical Experiences with Flow-Induced Vibrations. IUTAM/IAHR Symposium Karlsruhe 1979. Berlin-Heidelberg-New York, Springer-Verlag 1980. XXII, 849 S., 429 Abb., 217 Charts, DM 140,–. US $ 82.60. ISBN 3-540-10314-7

Es ist das erklärte Ziel der Veranstalter, die praktischen Aspekte der strömungsinduzierten Schwingungen in den Vordergrund zu stellen und insbesondere bewährte Konstruktionskriterien hervorzuheben. Das kann als durchaus gelungen bezeichnet werden. Die führenden Themen des Symposiums sind praktische Erfahrungen einschließlich Informationen über Versagen, Ermüdung und Verschleiß und, wo es möglich war, auch über entsprechende Abhilfemaßnahmen. Der Inhalt des Berichtsbandes ist gegliedert in:

- Reaktor- und Wärmetauscher-Baugruppen,
- hydraulische Maschinen und Ausrüstungen,
- hydraulische Baugruppen,
- Schiffsbauteile,
- meereswellenbeeinflußte Konstruktionen,
- Brückendecks, -balken und -kabel,
- Hochbauten,
- Vorhersage und Modellierung von strömungsinduzierten Schwingungen (Workshop),
- ungelöste Probleme (Einführungsvorträge einer Podiumsdiskussion).

Die aufgeführten Beiträge sind nicht nur für den Praktiker interessant und nützlich, sondern dürften auch dem theoretisch und experimentell arbeitenden Wissenschaftler manche Anregung geben. Das betrifft vor allem die Vorträge, in denen theoretische Ansätze und Experiment gegenübergestellt werden. Wenn man aus der Fülle der 102 gedruckten Beiträge überhaupt einige hervorheben sollte, so sind es die Übersichtsvorträge und darunter der Beitrag von M. PAÏDOUSSIS über strömungsinduzierte Schwingungen in Kernreaktoren und Wärmetauschern, der allein 52 Fälle anhand von Konstruktionsskizzen und Diagrammen darstellt. Besonders wertvoll sind wie in vielen anderen Beiträgen auch Hinweise und Erfahrungen zur Beseitigung der Schwingungen oder ihrer schädlichen Wirkungen.

Magdeburg U. FISCHER

Lippmann, H., Mechanik des Plastischen Fließens, Grundlagen und technische Anwendungen. Berlin-Heidelberg-New York, Springer-Verlag 1980. X, 358 S., 129 Abb., DM 148,–. US $ 87.40. ISBN 3-540-09925-5

Dieses Buch, das ursprünglich als Fortsetzung des 1. Bandes von H. LIPPMANN und O. MAHRENHOLTZ „Plastomechanik der Umformung metallischer Werkstoffe" gedacht war, erscheint nun als unabhängiges, in sich geschlossenes Werk über Plastizitätstheorie und ihre Anwendungen. Neben den umformtechnischen Problemen, die nach wie vor im Mittelpunkt stehen und mehr als die Hälfte des Inhaltes ausmachen, sind jetzt im Kapitel 1 noch Stabwerke sowie die Grundlagen des plastischen Kontinuums (Fließgesetze, Extremalsätze) und im Kapitel 2 einige geschlossene Lösungen (Torsion von Stäben, Blechbiegung, axialsymmetrische Blechumformung) enthalten. Dadurch gewinnt das Werk an theoretischem Gehalt. Zum besseren Verständnis wird im Anhang noch eine kurze Einführung in die Matrizenalgebra, die Grundbegriffe der Kontinuumsmechanik und spezielle (hyperbolische) Systeme partieller Differentialgleichungen gegeben. Damit ist das Buch zum Selbststudium für Ingenieure und Naturwissenschaftler, die sich in das Gebiet einarbeiten wollen, gut geeignet. Im Kapitel 3, elementare Plastizitätstheorie genannt, geht es um elementare Umformprozesse des Stauchens und Schmiedens ebener und zylindrischer Blöcke, des Drahtziehens und des Walzens von Blech. Ergänzt werden diese Ausführungen durch Probleme des Gebirgsschlags. Das 4. Kapitel überschreibt der Autor mit „Schrankenverfahren und verwandte Methoden". Hier werden Trennvorgänge und Zug-Druck-Umformungen von Metallen behandelt. Traglastprobleme bei körnigem Material zu zwei Beispielen der Bodenmechanik runden das Ganze ab. Offensichtlich zu kurz werden zum Schluß dieses 4. Kapitels die Numerischen Methoden (FEM) auf 6 Seiten abgehandelt. Wenn man den Umfang der heutigen Forschungen auf diesem Gebiet auch in Anwendung auf die Umformtechnik bedenkt, muß man hierzu in einem solchen Werk eine umfangreichere und auch positivere Bilanz zu diesen Methoden erwarten. Im 5. und letzten Kapitel werden die Charakteristiken-Verfahren ausführlich beschrieben. Sehr gut fügt sich dabei der Abschnitt „Plastokinetik des Einzelstabes" zu Beginn des Kapitels mit einer Anwendung auf das Hochgeschwindigkeitsstauchen ein. Im Abschnitt „Ebenes Fließen" werden die in großem Umfang vorhandenen konventionellen Gleitlinienlösungen dargelegt. Auch hier zweifelt der Autor erneut an der Verläßlichkeit der numerischen Methoden, speziell der FEM, denen er mangelnde Anschaulichkeit und Nachprüfbarkeit vorwirft. Dieser Vorwurf erscheint jedoch auf Basis des einen hier wieder angeführten Beispiels nach Bild 4.21 für das Stauchen zylindrischer Proben nicht genug fundiert. Den Schluß des 5. Kapitels bilden Probleme des axialsymmetrischen Fließens, die mit der Gleitlinien- bzw. Hauptlinientheorie gelöst werden können, sofern sie auf hyperbolische Differentialgleichungen führen. Am Ende des Buches ist ein ausführliches Literaturverzeichnis mit über 600 Quellenangaben zu finden, das die gesamte einschlägige Literatur enthält. Allerdings sind sowjetische Arbeiten nur gering vertreten. Insgesamt bietet das Buch sowohl dem praktisch als auch dem theoretisch tätigen Ingenieur einen sehr guten Überblick über das Gebiet der Plastizitätstheorie mit ihren vielseitigen technischen Anwendungen vor allem auf umformtechnische Probleme, wobei die theoretische Durchdringung das Gesamtwerk abrundet.

Dresden G. LANDGRAF

Lüst, R., Hydrodynamik, Mannheim-Wien-Zürich, B. I. – Wissenschaftsverlag 1978, 234 S., ISBN 3-411-01540-3

Diese einführende Darstellung, entstanden aus einem Vorlesungsmanuskript, wendet sich an mittlere Semester. Die einleitenden Kapitel stellen die Grundgleichungen der Fluidmechanik bereit, im Hauptteil werden viskose bzw. ideale Medien sowie Oberflächen- und Schallwellen behandelt, zwei kurze Schlußkapitel schneiden die Themen Stoßwellen bzw. Turbulenz an. — Das Buch enthält keine Aufgaben, aber genügend Beispiele und ausführliche mathematische bzw. physikalische Interpretationen des jeweiligen Sachverhaltes. Die Korrekturen hätten sorgfältiger gelesen werden können, man sehe z. B. nur das Durcheinander von φ und Φ in Abschnitt 1.9. — Insgesamt ist dieses Buch eine gute Einführung.

Rostock M. HEINRICH

Bosznay, Á. (ed.), Bracketing of Eigenfrequencies of Continuous Structures, Euromech Colloquium 112, Mátrafüred (Hungary), 1979, Budapest, Akadémiai Kiadó 1980, 669 S., Ft 480,–. ISBN 963-05-2625-5

This volume contains the Proceedings of the Euromech Colloquium No. 112, held in Mátrafüred (Hungary) on February 21 to 23, 1979. The Colloquium was organized by the Research Group of Technical Mechanics of the Hungarian Academy of Sciences. The main objective of the Colloqium was to bring together more closely mathematicians and engineers interested in mechanics and to discuss problems of advanced engineering mechanics. The organizing committee succeeded in gathering of outstanding scientists all over the world (not just from Europe) who are involved in up-to-date theoretical and applied mechanics, to name at least some of them: N. BAZLEY, G. FICHERA, D. W. FOX, R. UHRIG, W. H. WITTRICK, S. NEMAT-NASSER, V. M. KORNEV, Á. BOSZNAY, B. L. KARIHALOO and many others.

There are 38 papers printed in the Proceedings, which basically deal with problems of modifications, refinements, ramifications, and new applications of the Poincaré-Rayleigh-Ritz method, further of the Weinstein-Aronszajn-Bazley-Fox method for assessment of the lower and upper bounds, of Fichera's method of orthogonal invariants. A great number of papers deals with variational characterization of eigenvalues and optimization problems in connection with eigendata of vibrating mechanical systems (linear and non-linear). Some new methods which assure the determination of all eigenfrequencies in a given interval were treated as well and special eigenvalue problems in composite materials and reinforced plastics were also dealt with.

In the reviewer's view the meeting was a great success and one feels that a regular series of similar uninformal meetings on the „bracketing techniques of eigenvalues" should be now inaugurated where various aspects of relevant problems could be discussed and which would include all branches of mechanics. In conclusion it may be said that this volume will find a good response among specialists and will be of undisputed value to both, mathematicians and engineers who really feel that some "refinements,, are still needed in calculations of their eigendata.

Bratislava Š. Markuš

Brandt, S. / Dahmen, H. D., Physik. Eine Einführung in Experiment und Theorie. Bd. 2: Elektrodynamik. Berlin-Heidelberg-New York, Springer-Verlag 1980. XVIII, 586 S., 219 Abb., DM 55,—. US $ 32.50. ISBN 3-540-09947-6

Der vorliegende Elektrodynamik-Band dieses neuen Lehrbuches der Physik setzt in gelungener Weise die Linie des bereits gut eingeführten Mechanik-Bandes (1977) fort. Die wichtigsten Experimente werden wieder mit quantitativen Auswertungen vorgestellt, z. B. in Form von Photos und Oszillogrammen, die Zeichnungen sind mit großer Sorgfalt und z. T. mit Hilfe des Computers ausgeführt und vermitteln maximale Informationen. Da auch der theoretische Teil hohen Ansprüchen genügt, verdient dieses Gemeinschaftswerk eines experimentellen und eines theoretischen Physikers große Aufmerksamkeit. Der Stoff wird in 3 Zügen dargestellt, die verschieden orientiert sind und teilweise parallel laufen. Im Hauptzug wird die Elektrodynamik aus Grundlagenexperimenten heraus induktiv entwickelt, wobei die Darstellung mit den elektrischen Feldern statischer Ladungen und den magnetischen Feldern stationärer Ströme beginnt und in den Maxwellgleichungen sowie der Theorie elektromagnetischer Wellen gipfelt. Der zweite Zug spricht besonders den Theoretiker an, indem er die Symmetrien der raumzeitlichen Transformationen in den Mittelpunkt stellt. Die Maxwellgleichungen werden auf einem eleganten Wege als relativistische Verallgemeinerungen der Feldgleichungen der Elektrostatik erhalten. Der dritte Zug konzentriert sich auf die Zusammenhänge zwischen elektrischen und magnetischen Materialeigenschaften und dem mikroskopischen Bau der Materie. Die mikroskopische Begründung der Gesetze der Felder und des Ladungstransports in Stoffen wird gegeben. Das Bändermodell der Festkörper wird auf der Basis der Fermistatistik dargestellt, und schließlich wird die Physik der wichtigsten Bauelemente der Elektronik wie Diode und Transistor entwickelt. Die Stärken und Schwächen des Buches resultieren mehr oder weniger aus der sehr weiten Zielstellung. Sowohl der Student als auch der ausgebildete Physiker werden die Vielseitigkeit dieser experimentell und theoretisch tiefgründigen und umfassenden Darstellung sehr begrüßen. Andererseits ist ein Hochschullehrbuch der Elektrodynamik mit rund 600 Seiten eigentlich zu lang; von manchen Details hätte man vielleicht doch absehen können, das betrifft auch die Anhänge. Zu den unbestrittenen Verdiensten der neuen Lehrbuchreihe zählt sicher, daß die Autoren neue methodische Wege mit großem didaktischen Geschick eingeschlagen haben, was dem Werk einen breiten Leserkreis sichern sollte.

Berlin W. Ebeling

Lücker, R., Grundlagen digitaler Filter. Einführung in die Theorie linearer zeitdiskreter Systeme und Netzwerke. Nachrichtentechnik 7. Berlin-Heidelberg-New-York, Springer-Verlag 1980. X, 255 S., 96 Abb., DM 58, –. US $ 34.30. ISBN 3-540-10198-5

Die vorliegende Monografie, entstanden aus dem Skriptum einer Vorlesung an der TH Darmstadt, bietet eine Einführung in die Theorie digitaler Filter. Vorausgesetzt werden mathematische Kenntnisse, wie sie allgemein beim Studium der Ingenieurwissenschaften vermittelt werden; weitergehende Kenntnisse auf dem Gebiet der Theorie der komplexen Funktionen sind zum besseren Verständnis des Stoffes nützlich.

Der Inhalt gliedert sich in 3 Hauptteile: Im ersten werden systemtheoretische Grundlagen und Begriffe vermittelt (zeitdiskrete, LTI-, FIR-, IIR-Systeme, z-Transformation, Übertragungsfunktion usw.). Der zweite, umfangreichste Teil ist der Theorie zeitdiskreter Netzwerke gewidmet, Es werden verschiedene kanonische und nichtkanonische Darstellungsweisen von Systemen mit ein und mehreren Ein- und Ausgängen beschrieben. Die mathematische Behandlung erfolgt mittels Signalflußgraphen und deren Übertragungs- und Zustandsgleichungen. Die Analyse zeitdiskreter Netzwerke erfolgt durch Berechnung der Zustandsmatrizen. Numerische Frequenzgangs- und Stabilitätsuntersuchungen schließen sich an. Die strukturunabhängige Synthese von zeitdiskreten Netzwerken wird kurz dargestellt und anhand von Beispielen demonstriert. Der dritte Teil des Buches befaßt sich speziell mit dem Entwurf zeitdiskreter IIR-Systeme: Kurz dargelegt wird der Systementwurf nach vorgegebener Impulsantwort; eine ausführliche Zusammenstellung verschiedener Entwurfsmethoden nach Vorschriften im Frequenzbereich (selektive Filter) schließt sich an. Neben Frequenztransformationen werden auch die wichtigsten Standardlösungen zum Entwurf zeitkontinuierlicher Filter angegeben. Eine kurze Darstellung einer Methode zur Berechnung von sog. „Wellendigitalfiltern" schließt diesen Teil ab. — Praktisch als Anhang folgt noch eine kurze Beschreibung des Entwurfs linearphasiger FIR-Systeme mit dem bekannten Verfahren der Fourier-Transformation und Fensterung. Hier wäre eine ausführlichere Darstellung wünschenswert.

Das Buch wird seinem einführenden Charakter, ohne Kompromisse bezüglich mathematischer Strenge und Geschlossenheit in der Darstellung einzugehen, auch durch eine große Anzahl sorgfältig ausgewählter und ausführlich erläuterter Beispiele gerecht. Eine Vielzahl von Abbildungen von Netzwerken, Frequenzgängen usw. tragen zum besseren Verständnis des Stoffes bei. Ferner werden Hinweise zur numerischen Behandlung der Probleme gegeben, im Anhang sind FORTRAN-Programme zur Berechnung vollständger elliptischer Integrale und elliptischer Funktionen wiedergegeben. Ein umfangreiches Literaturverzeichnis schließt dieses sehr verständlich geschriebene Einführungswerk ab.

Berlin U. Schmucker

Wuytack, L. (Ed.), Padé Approximation and Its Applications. Proceedings, Antwerp, Belgium, 1979. Lecture Notes in Mathematics 765. Berlin-Heidelberg-New York, Springer-Verlag 1979. VI, 392 S., 14 Abb., DM 35,50. US $ 19.90. ISBN 3-540-09717-1

Die Konferenz in Antwerpen 1979 ist als Fortsetzung zweier 1977 und 1978 von Claude Brezinski an der Universität von Lille (Frankreich) organisierter anzusehen. In den Konferenzbericht wurden Beiträge zu folgenden Themen aufgenommen:

1. Mathematische Aspekte der Padé-Approximation und ihrer Verallgemeinerungen (Padé-Hermite Approximation, Padé-Chebychev-Approximation, Abstrakte Padé-Approximation in der Operatorentheorie)
2. Berechnung von Padé-Approximierenden
3. Anwendungen der Padé-Approximation z. B. bei Runge-Kutta-Verfahren und Prädiktor-Korrektor-Verfahren für nichtlineare Volterra-Integralgleichungen zweiter Art).

Hervorzuheben sind zwei Bibliographien. Die von C. Brezinski ergänzt seine früher gegebene Literaturübersicht über Padé-Approximationen. Eine zweite, von L. Wuytack zusammengestellte und mit Kurzreferaten versehene, beschränkt sich auf wesentliche Arbeiten auf diesem Gebiet.

Berlin W. Zielas

Sattinger, D. H., Group Theoretic Methods in Bifurcation Theory. Lecture Notes in Mathematics 762. Berlin-Heidelberg-New York, Springer-Verlag 1979. V, 241 S., 14 Abb., DM 25,—. US $ 14.00. ISBN 3-540-09715-5

Das Verhältnis von Bifurkation und Symmetrie(brechung) ist schon seit langem ein besonders reizvolles Thema, insbesondere der mathematischen Physik. Zum Beispiel hat der Zusammenhang von Symmetriebrechung durch Störung und der durch diese Störung bedingten (qualitativen) Veränderung des

Spektrums (Aufspaltung von Eigenwerten) seinerzeit in der gruppentheoretischen Behandlung der Atomspektren (E. WIGNER, Gruppentheorie und ihre Anwendungen auf die Quantenmechanik der Atome, Berlin 1931) einen großen Triumph gefeiert. Der Verfasser der vorliegenden Schrift unternimmt es, die der Beziehung zwischen Symmetriebrechung und Bifurkation zugrunde liegende Kombination der Theorie der Gruppendarstellungen und der Bifurkationstheorie zusammenhängend darzustellen. Zunächst (Kap. I) werden einige physikalische Beispiele für Bifurkationsprobleme erläutert (Hydromechanik, Astrophysik, Elastizitätstheorie, Elementarteilchentheorie). Danach (Kap. II) werden einige mathematische Hilfsmittel (aus linearer und nichtlinearer Funktionalanalysis und der Theorie elliptischer und parabolischer Differentialgleichungen) rekapituliert. Die beiden nächsten Kapitel (III und IV) sind dann den grundlegenden Tatsachen und Methoden der Bifurkationstheorie gewidmet (u. a. Reduktion des Bifurkationsproblems auf ein endlichdimensionales, rein algebraisches Problem, d. h. Herleitung der Bifurkationsgleichung durch die sog. Ljapunov-Schmidt-Prozedur; Newtonsches Diagramm; Zusammenhang von Stabilität der Bifurkationslösungen und Eigenschaften der Funktionaldeterminante der reduzierten Bifurkationsgleichung (unter Benutzung der Störungstheorie isolierter Eigenwerte endlicher Vielfachheit)). Die beiden letzten Abschnitte von Kap. IV (Kovarianz bzw. Tensorcharakter der Bifurkationsgleichung) leiten über zur Anwendung gruppentheoretischer Methoden (verbindendes Element ist die Theorie der Tensorprodukte von Gruppendarstellungen). Kap. V enthält die Elemente der Darstellungstheorie von endlichen und kompakten Gruppen (im wesentlichen für endlich-dimensionale Darstellungen), einschließlich der Theorie der Charaktere und Tensorprodukte (speziell werden Lie-Gruppen betrachtet). Das Problem der Aufstellung von Bifurkationsgleichungen gegebener Symmetrie wird behandelt (für SO(3)-Symmetrie). Das letzte Kap. VI enthält Anwendungen der gruppentheoretischen Methoden auf einige Probleme der angewandten Mathematik (u. a. Brechung Euklidischer Invarianz in der Ebene (Bénard-Problem); Konstruktion von Bifurkationsgleichungen für gegebene Gitter; Bifurkation rotationssymmetrsicher Zustände; Variationsmethoden). Ein Anhang (von P. OLVER) über algebraische Methoden zur Herleitung von Symmetriegruppen für Differentialgleichungen beschließt die Schrift.

Berlin H. BAUMGÄRTEL

In der DDD beziehbar im Rahmen des Kontinents.

EINGEGANGENE BÜCHER

Bei der Schriftleitung sind folgende Bücher eingegangen (ausführliche Besprechung bleibt vorbehalten).

Baldock, G. R. / **Bridgeman, T.,** The Mathematical Theory of Wave Motion. New York-Chichester-Brisbane-Toronto, John Wiley & Sons and Ellis Horwood Limited, Chichester 1981. 261 S., £ 19.50. ISBN 0-470-27113-2

Gaus, W. (Hrsg.), Ausbildung in Medizinischer Dokumentation, Statistik und Datenverarbeitung. Proceedings, Neu-Ulm, 1979. Medizinische Informatik und Statistik 25. Berlin-Heidelberg-New York, Springer-Verlag 1981. X, 122 S., DM 25,–. US $ 14.80. ISBN 3-540-10280-9

Victor, N. / **Lehmacher, W.** / **Eimeren, W. van** (Hrsg.), Explorative Datenanalyse, Frühjahrstagung, München, 1980. Medizinische Informatik und Statistik 26. Berlin-Heidelberg-New York, Springer-Verlag 1980. V, 211 S., DM 35,–. US $ 20.70. ISBN 3-540-10281-7

Davis, P. J. / **Hersh, R.,** The Mathematical Experience. Boston-Basel-Stuttgart, Birkhäuser Verlag 1981. XIX, 440 S., sFr. 52.–. ISBN 3-7643-3018-X

Blaquière, A. / **Fer, F.** / **Marzollo, A.** (eds.), Dynamical Systems and Microphysics. CISM Courses and Lectures 261. Wien-New York, Springer-Verlag 1980. IX, 412 S., 33 Abb., DM 66,–. S 472,–. US $ 39.00. ISBN 3-211-81533-3

Fischer, U. / **Stephan, W.,** Mechanische Schwingungen. Leipzig, VEB Fachbuchverlag 1980. 332 S., M 22,–. BN 546-533-8

Lewis, R. W. / **Morgan, K.** / **Zienkiewicz, O. C.** (eds.), Numerical Methods in Heat Transfer. Chichester-New York-Brisbane-Toronto, John Wiley & Sons 1981. XV, 536 S., £ 28.00. ISBN 0-471-27803-3

Barlow, R. E. / **Proschan, F.,** Statistische Theorie der Zuverlässigkeit. 2. Aufl., Berlin, Akademie-Verlag 1981. X, 250 S., 58 Abb., M 48,–. BN 762 232 5 (6438)

Everitt, W. N. (ed.), Ordinary and Partial Differential Equations, Proceedings, Dundee, Scotland 1978. Lecture Notes in Mathematics 827. Berlin-Heidelberg-New York, Springer-Verlag 1980. XVI, 271 S., 6 Abb., DM 29,–. US $ 17.20. ISBN 3-540-10252-3

Drazin, P. / **Reid, W.,** Hydrodynamic Stability Cambridge, Cambridge University Press 1981. XI, 525 S. £ 35.00 A H/C. SBN 0-521-22798-4

Hawking, S. W. / **Rocek, M.** (eds.), Superspace and Supergravity, Cambridge, Cambridge University Press 1981. XII, 527 S., £ 24.00 A H/C. ISBN 0-521-23908-7

Noltemeier, H., Informatik I, Einführung in Algorithmen und Berechenbarkeit. München-Wien, Carl Hanser Verlag 1981. 148 S., DM 32,–. ISBN 3-446-13388-7

Flügge, S., Mathematische Methoden der Physik II, Geometrie und Algebra. Berlin-Heidelberg-New York, Springer-Verlag 1980. VI, 174 S., 19 Abb., DM 38,–. US $ 22.50. ISBN 3-540-10062-8

Herman, G. T. / **Natterer, F.** (eds.), Mathematical Aspects of Computerized Tomography. Proceedings, Oberwolfach 1980. Lecture Notes in Medical Informatics 8. Berlin-Heidelberg-New York, Springer-Verlag 1981. VII, 309 S., DM 47.50. US $ 28.10. ISBN 3-540-10277-9

Ciarlet, G. / **Rabier, P.,** Les équations de von Kármán. Lecture Notes in Mathematics 826. Berlin-Heidelberg-New York, Springer-Verlag 1980. VI, 181 S., 15 Abb., DM 21,50. US $ 12.70. ISBN 3-540-10248-5

Schauer, H. / **Barta, G.,** Methoden der Programmerstellung für Tisch- und Taschenrechner. Grundlagen, Anwendungen, Grenzen. Wien-New-York, Springer-Verlag 1979, X, 190 S., 134 Abb., S 269,–. DM 39,–. US $ 21.50. ISBN 3-211-81476-0

Parkus, H. (ed.), Electromagnetic Interactions in Elastic Solids. CISM Courses and Lectures 257. Wien-New York, Springer-Verlag 1979. VI, 425 S., 15 Abb., S 486,–. DM 68,–. U$ S 38.10. ISBN 3-211-81509-0

Paul, B., Kinematics and Dynamics of Planar Machinery. Englewood Cliffs, Prentice-Hall, Inc., 1979. XVII, 670 S., US $ 44.50. ISBN 0-13-516062-6

Müller, P. C. / Schiehlen, W. O., Forced Linear Vibrations. CISM Courses and Lectures 172. Wien-New York, Springer-Verlag 1977. 146 S., 35 Abb., DM 36,–. S 248,–. US $ 19.80. ISBN 3-211-81487-6

Walter, E., Biomathematik für Mediziner. Stuttgart, 2. Aufl., B. G. Teubner 1980. 206 S., DM 19.80. ISBN 3-519-12049-6

Morecki, A.(ed.), Biomechanics of Motion. Wien-New York, Springer-Verlag 1980. XVI, 217 S., 74 Abb., DM 49,–. S 350,–. US $ 29.00. ISBN 3-211-82611-9

Guessarian, I., Algebraic Semantics. Lecture Notes in Computer Science 99. Berlin-Heidelberg-New York, Springer-Verlag 1981. 158 S., DM 21,50. US $ 11.30. ISBN 3-540-10284-1

Noltemeier, H. (ed.), Graphtheoretic Concepts in Computer Science. Proceedings, Bad Honnef, 1980. Lecture Notes in Computer Science 100. Berlin-Heidelberg-New York, Springer-Verlag 1981. X, 403 S., DM 41,50. US $ 21.80. ISBN 3-540-10291-4

Thayse, A., Boolean Calculus of Differences. Lecture Notes in Computer Science 101. Berlin-Heidelberg-New York. Springer-Verlag 1981. VII, 144 S., DM 18,–. US $ 9.50. ISBN 3-540-10286-8

Davenport, J. H., On the Integration of Algebraic Functions. Lecture Notes in Computer Scince 102. Berlin-Heidelberg-New York, Springer-Verlag 1981. II, 197 S., DM 21,50. US $ 11.30. ISBN 3-540-10290-6

Ledgard, H. / Singer, A. / Whiteside, J., Directions in Human Factors for Interactive Systems. Lecture Notes in Computer Science 103. Berlin-Heidelberg-New York, Springer-Verlag 1981. VI, 190 S., DM 21,50. US $ 10.30. ISBN 3-540-10574-3

Eisenbach, S. / Sadler, C., PASCAL for Programmers. Berlin-Heidelberg-New York, Springer-Verlag 1981. IX, 201 S., 78 Abb., DM 19,80. US $ 11.70. ISBN 3-540-10473-9

Schüßler, H. W., Netzwerke, Signale und Systeme. Band 1: Systemtheorie linearer elektrischer Netzwerke. Berlin-Heidelberg-New York, Springer-Verlag 1981. XI, 382 S., 210 Abb., DM 54,–. US $ 28.40. ISBN 3-540-10524-7

Elsner, R., Nichtlineare Schaltungen. Nachrichtentechnik 8. Berlin-Heidelberg-New York, Springer-Verlag 1981. IX, 136 S., 113 Abb., DM 54,–. US $ 25.70. ISBN 3-540-10477-1

Leipholz, H. H. E. (ed.), Stability of Elastic Structures. CISM Courses and Lectures 238. Wien-New York, Springer-Verlag 1978. VII, 291 S., 52 Abb., DM 48,–. S 332,– ISBN 3-211-81473-6

Wagner, M., Elemente der Theoretischen Physik 1: Klassische Mechanik, Quantenmechanik. 2. Auflage. Braunschweig-Wiesbaden, Vieweg 1980. 200 S., DM 14,80. ISBN 3-528-17221-5

Berg, H. K. / Giloi, W. K. (eds.). The Use of Formal Specification of Software. June 1979, Berlin. Informatik-Fachberichte 36. Berlin-Heidelberg-New York, Springer-Verlag 1980. V, 388 S., DM 42,–. US $ 24.80. ISBN 3-540-10442-9

Wagner, M., Elemente der Theoretischen Physik 2: Felder und Wellen, Klassische und statistische Thermodynamik. Reinbek bei Hamburg, Rowohlt Taschenbuchverlag GmbH 1977. 255 S., 66 Abb., DM 16,80. ISBN 3-499-27030-7

Hughes, P. / Brecht, G., Die Scheinwelt des Paradoxons. Eine kommentierte Anthologie in Wort und Bild. Braunschweig, Vieweg 1978. 120 S., ISBN 3-528-08379-4.

Ogilvy, C. S., Mathematische Leckerbissen. 2. Auflage, Braunschweig-Wiesbaden, Vieweg 1980. 111 S., 39 Abb., ISBN 3-528-18281-4

Ebert, F., Strömung nicht-newtonscher Medien. Braunschweig, Vieweg 1980. X, 190 S., DM 29,80. ISBN 3-528-03339-8

Lampson, B. W. / Paul, M. / Siegert, H. J. (eds.), Distributed Systems – Architecture and Implementation. An Advanced Course. Lecture Notes in Computer Science 105. Berlin-Heidelberg-New York, Springer-Verlag 1981. XIII, 510 S., DM 49,–. US $ 23.40. ISBN 3-540-10571-9.

The Programming Language Ada, Reference Manual. Proposed Standard Document, United States Department of Defense. Lecture Notes in Computer Science 106. Berlin-Heidelberg-New York, Springer-Verlag 1981. X, 243 S., DM 16,50. US $ 7.90. ISBN 3-540-10693-6

Archetti, F. /Cugiani, M. (eds.), Numerical Techniques for Stochastic Systems. Conference, Gargnano, Italy, 1979. Amsterdam-New York, North-Holland Publ. Co. 1980. VIII, 406 S., US $ 53.75. Dfl. 110.00. ISBN 0-444-86000-2

Deussen, P. (ed.), Theoretical Computer Sciencet. 5 th GI-Conference, Karlsruhe 1981. Lecture Notes in Computer Science 104. Berlin-Heidelberg-New York, Springer-Verlag 1981. VII, 261 S., DM 29,50. US $ 14.10. ISBN 3-540-10576-X

Dokumentation Rheologie. 1980. Nr. 46. Bundesanstalt für Materialprüfung, Fachgruppe Rheologie und Trbologie und DRG, D-1000 Berlin 45, Unter den Eichen 87. 863 S., ISSN 0340-8388

Rimrott, F. P. J. / Tabarrok, B. (ed.), Theoretical and Applied Mechanics. Proceedings, IUTAM, Toronto 1980. Amsterdam-New York. North-Holland Publ. Co. 1981. XXXII, 458 S., US $ 78.00. Dfl. 160.00. ISBN 0-444-85411-8

Vöge, K. H. (Hrsg.), Telekommunikation für Bildung und Ausbildung. Berlin-Heidelberg-New York, Springer-Verlag 1981. VII, 108 S., DM 30,–. US $ 14.30. ISBN 3-540-10645-6

Unbehauen, R., Elektrische Netzwerke. Eine Einführung in die Analyse. 2. Auflage. Berlin-Heidelberg-New York, Springer-Verlag 1981. XIII, 431 S., 314 Abb., DM 54,–. US $ 25.70. ISBN 3-540-10543-3

Unbehauen, R. / Hohneker, W., Elektrische Netzwerke. Aufgaben. Ausführlich durchgerechnete und illustrierte Aufgaben zur Netzwerkanalyse mit Lösungen. Berlin-Heidelberg-New York, Springer-Verlag 1981. 360 S., 87 Abb., DM 54,–. US $ 25.70. ISBN 3-540-10542-5

Hult, J. / Lemaitre, J. (eds.), Physical Non-Linearities in Structural Analysis. IUTAM-Symposium Senlis, France, 1980. Berlin-Heidelberg-New York, Springer-Verlag 1981. XI, 287 S., 109 Abb., DM 64,–. US $ 30.50. ISBN 3-540-10544-1

Strubecker, K., Einführung in die Höhere Mathematik mit Berücksichtigung ihrer Anwendungen auf Geometrie, Physik, Naturwissenschaften und Technik. Band 3: Integralrechnung einer reellen Veränderlichen. München-Wien, R. Oldenbourg Verlag 1980. 822 S., 256 Abb., DM 88,–. ISBN 3-486-22421-2

Mertens, B. (Hrsg.), Messung, Modellierung und Bewertung von Rechensystemen. GI-NTG Fachtagung, Jülich 1981. Informatik-Fachberichte 41. Berlin-Heidelberg-New York, Springer-Verlag 1981. VIII, 368 S., DM 42,–. US $ 22.10. ISBN 3-540-10619-7

Goos, G. (Hrsg.), Werkzeuge der Programmiertechnik. GI-Arbeitstagung, Karlsruhe 1981. Informatik-Fachberichte 43. Berlin-Heidelberg-New York, Springer-Verlag 1981. VI, 262 S., DM 33,–. US $ 15.70. ISBN 3-540-10725-8

Baumann, R. (Hrsg.), Fachtagung Prozeßrechner 1981, München 1981. Informatik-Fachberichte 39. Berlin-Heidelberg-New York, Springer-Verlag 1981. XVI, 476 S., DM 51,–. US $ 26.80. ISBN 3-540-10617-0

Schindler, S. / Schröder, J. C. W. (Hrsg.), Kommunikation in verteilten Systemen. Fachtagung der GI, Berlin 1981. Informatik-Fachberichte 40. Berlin-Heidelberg-New York, Springer-Verlag 1981. IX, 459 S., DM 51,–. US $ 30.10. ISBN 3-540-10618-9

Doebner, H. D. (ed.), Differential Geometric Methods in Mathematical Physics. Proceedings, Clausthal, 1978. Lecture Notes in Physics 139. Berlin-Heidelberg-New York, Springer-Verlag 1981. VII, 329 S., DM 38,–. US $ 18.10. ISBN 3-540-10578-6

Kramer, P. / Saraceno, M., Geometry of the Time-Dependent Variational Principle in Quantum Mechanics. Lecture Notes in Physics 140. Berlin-Heidelberg-New York, Springer-Verlag 1981. IV, 98 S., DM 18,–. US $ 8.60. ISBN 3-540-10579-4

Reynolds, W. C. / MacCormack, R. W. (eds.), Seventh International Conference on Numerical Methods in Fluid Dynamics. Proceedings, Stanford 1980. Lecture Notes in Physics 141. Berlin-Heidelberg-New York, Springer-Verlag 1981. VIII, 485 S., DM 52,–. US $ 24.80. ISBN 3-540-10694-4

Eichmeier, J., Moderne Vakuumelektronik. Grundlagen, Bauelemente, Technologie. Berlin-Heidelberg-New York, Springer-Verlag 1981. XI, 412 S., 430 Abb., DM 178,–. US $ 84.80. ISBN 3-540-10212-4

Nussbaumer, H. J., Fast Fourier Transform and Convolution Algorithms. Springer Series in Information Sciences 2. Berlin-Heidelberg-New York, Springer-Verlag 1981. X, 248 S., 34 Abb., DM 69,–. US $ 40.80. ISBN 3-540-10159-4

Eckhaus, W. / van Harten, A., The Inverse Scattering Transformation and the Theory of Solitons. An Introduction. North-Holland Publ. Co. 1981. XI. 224 S., US $ 31.75. Dfl. 65.00. ISBN 0-444-86166-1

Hirschel, E. H. / Kordulla, W., Shear Flow in Surface-Oriented Coordinates. Notes on Numerical Fluid Mechanics 4. Braunschweig-Wiesbaden, Vieweg 1981. X, 266 S., ISBN 3-528-08078-7

Schuon, E. / Wolf, H., Nachrichten-Meßtechnik. Nachrichtentechnik 9. Berlin-Heidelberg-New York, Springer-Verlag 1981. XI, 271 S., 155 Abb., DM 52,–. US $ 24.80. ISBN 3-540-10637-5

Pflüger, A., Elementare Schalenstatik. 5. Auflage. Berlin-Heidelberg-New York, Springer-Verlag 1981, VIII, 128 S., 59 Abb., DM 52,–. US $ 24.80. ISBN 3-540-10320-1

Jimenez, J. (ed.), The Role of Coherent Structures in Modelling Turbulence and Mixing. Proceedings, Madrid 1980. Lecture Notes in Physics 136. Berlin-Heidelberg-New York, Springer-Verlag 1981. XIII, 393 S., DM 47,–. US $ 24.70. ISBN 3-540-10289-2

Katok, A. (ed.), Ergodic Theory and Dynamic Systems I. Proceedings, Special Year, Maryland 1979–80. Progress in Mathematics 10. Boston-Basel Stuttgart, Birkhäuser Verlag 1981. XI, 333 S., ISBN 3-7643-3036-8

Österle, H., Entwurf betrieblicher Informationssysteme. Computer Monographien 16. München-Wien, Carl Hanser Verlag 1981. 301 S., DM 38,–. ISBN 3-446-13359-3

Mayr, R., Automatisierte Netzgenerierung für Finite-Element-Verfahren. Produktionstechnik Berlin 20. München-Wien, Carl Hanser Verlag 1981. 270 S., ISBN 3-446-13308-9

Diaz, J. / Ramos, I. (eds), Formalization of Programming Concepts. International Colloquium, Peniscola, Spain, 1981. Lecture Notes in Computer Science 107. Berlin-Heidelberg-New York, Springer-Verlag 1981. VII, 478 S., DM 45,50. US $ 21.70. ISBN 3-540-10699-5

Olle, T. W., Das Codasyl-Datenbankmodell. Berlin-Heidelberg-New York, Springer-Verlag 1981. XXIV, 389 S., DM 79,–, US $ 37.60. ISBN 3-540-10669-3 (Übersetzung aus dem Englischen)

Wunderlich, W. / Stein, E. / Bathe, K.-J. (eds.), Nonlinear Finite Element Analysis in Structural Mechanics. Proceedings of the Europe-U.S. Workshop, Bochum 1980. Berlin-Heidelberg-New York, Springer-Verlag 1981. XIII, 777 S., DM 78,–. US $ 37.20. ISBN 3-540-10582-4

Ropohl, G., Eine Systemtheorie der Technik. Zur Grundlegung der Allgemeinen Technologie. München-Wien, Carl Hanser Verlag 1979. 336 S., DM 44,–. ISBN 3-446-12801-8

Kreißig, R. / Drey, K.-D. / Naumann, J., Methoden der Plastizität. Anwendung auf Umformprobleme. München-Wien, Carl-Hanser Verlag 1980. 204 S., 80 Abb., DM 32,–. ISBN 3-446-13148-5

Kalide, W., Technische Strömungslehre. Eine Einführung. 5., erw. Auflage. Studienbücher der technischen Wissenschaften. München-Wien, Carl Hanser Verlag 1980. 252 S., 198 Abb., DM 32,–. ISBN 3-446-13092-6

Kalide, W., Aufgabensammlung zur technischen Strömungslehre. 3., erw. Auflage. Studienbücher der technischen Wissenschaften. München, Carl Hanser Verlag 1979. 155 S., DM 28,–. ISBN 3-446-12808-5

Schwalbe, K.-H., Bruchmechanik metallischer Werkstoffe. München-Wien, Carl Hanser Verlag 1980. 847 S., 500 Abb., DM 148,–. ISBN 3-446-12983-9

Meschkowski, H., Problemgeschichte der Mathematik II. Mannheim-Wien-Zürich, B.I.-Wissenschaftsverlag 1981. 235 S., DM 36,–. ISBN 3-411-01592-6

Landau, L. D. / Lifschitz, E. M., Lehrbuch der Theoretischen Physik I. Mechanik. 10., unveränderte Auflage. Berlin, Akademie-Verlag 1981. 231 S., M 16,–. BN 7602017 (5436/I)

Lauwerier, H. A., Mathematical Models of Epidemics. Mathematical Centre Tracts 138. Amsterdam, Mathematisch Centrum 1981. 125 S., Dfl. 15,75. ISBN 9061962161

Fahrmeir, L. / Kaufmann, H. / Ost, F., Stochastische Prozesse. Eine Einführung in Theorie und Anwendungen. München-Wien, Carl Hanser Verlag 1981. XII, 366 S., 44 Abb., DM 44,–. ISBN 3-446-13411-5

De Santo, J. A. / Sáenz, A. W. / Zachary, W. W. (eds.), Mathematical Methods and Applications of Scattering Theory. Proceedings, Washington, D. C. 1979. Lecture Notes in Physics 130. Berlin-Heidelberg-New York, Springer-Verlag 1980. XIII, 331 S., 76 Abb., DM 38,–. US $ 22.50. ISBN 3-540-10023-7

Gelman, I. W. / Mazja, W. G., Abschätzungen für Differentialoperatoren im Halbraum. Berlin, Akademie-Verlag 1981. 221 S., M 45,–. BN 7628145 (6581)

Bernasconi, J. / Schneider, T. (eds.), Physics in One Dimension. Solid-State Sciences 23. Berlin-Heidelberg-New York, Springer-Verlag 1981. IX, 368 S., 176 Abb., DM 65,–. US $ 34.20. ISBN 3-540-10586-7

Eilenberger, G., Solitons. Mathematical Methods for Physicists. Solid State Sciences 19. Berlin-Heidelberg-New York, Springer-Verlag 1981. VIII, 192 S., 31 Abb., DM 56,–. US $ 26.70. ISBN 3-540-10223-X

Hassard, B. D. / Kazarinoff, N. D. / Wan, Y.-H., Theory and Applications of Hopf Bifurcation. London Mathematical Society Lecture Note Series 41. Cambridge, Cambridge University Press 1981. 320 S., £ 15.00 A P/B. ISBN 0521231582

Ibach, H. / Lüth, H., Festkörperphysik. Eine Einführung in die Grundlagen. Berlin-Heidelberg-New York, Springer-Verlag 1981 IX, 238 S., 120 Abb., DM 54,–. US $ 28.40. ISBN 3-540-10454-2

Grenander, U., Regular Structures. Lectures in Pattern Theory, vol. 3. Applied Mathematical Sciences 33. Berlin-Heidelberg New York, Springer-Verlag 1981. VIII, 569 S., DM 53,–. US $ 25 30 ISBN 3-540-90560-X

Bratteli, O. / Robinson, D. W., Operator Algebras and Quantum Statistical Mechanics II Equilibrium States. Models in Quantum Mechanics. Texts and Monographs in Physics. Berlin-Heidelberg-New York, Springer-Verlag 1981. XI, 505 S, DM 86,–. US $ 41.00 ISBN 3-540-10381-3

The Chern Symposium 1979. Proceedings of the International Symposium on Differential Geometry in honor of S.-S. Chern, Berkely, California, 1979. Berlin-Heidelberg-New York, Springer-Verlag 1980. VII, 259 S., DM 56,–. US $ 33.10. ISBN 3-540-90537-5

Goldstine, H. H., A History of the Calculus of Variations from the 17th through the 19th Century. Studies in the History of Mathematics and Physical Sciences 5. Berlin-Heidelberg-New York, Springer-Verlag 1980. XVIII, 410 S., 66 Abb., DM 88,–. US $ 51.80. ISBN 3-540-90521-9

Ramm, A. G., Theory and Applications of Some New Classes of Integral Equations. Berlin Heidelberg-New York. Springer-Verlag 1980. XIII, 343 S, 2 Abb., DM 37,50. US $ 22.20. ISBN 3-540-90540-5

Gardiner, C. F., A First Course in Group Theory, Berlin-Heidelberg-New York, Springer-Verlag 1980. IX. 227 S., DM 28,–. US $ 16.60. ISBN 3-540-90545-6

Swinney, H. L. / Gollub, J. P. (eds.), Hydrodynamic Instabilities and the Transition to Turbulence. Topics in Applied Physics 45. Berlin-Heidelberg-New York, Springer-Verlag 1981. XII, 292 S, 81 Abb, DM 96,–. US $ 50.40. ISBN 3-540-10390-2

Goujet, C. / Nicolas, C., Mathématiques appliquées Proabilités initiation à la recherche opérationnelle. Paris, MASSON 1981 271 S, ISBN 2-225-74063-1

John, F., Plane Waves and Spherical Means, Applied to Partial Differential Equations. Reprint. Berlin-Heidelberg-New York, Springer-Verlag 1981. VIII, 172 S, 10 Abb., DM 42,–. US $ 20.00. ISBN 3-540-90565-0

Glowinski, R., Lectures on Numerical Methods for Non-Linear Variational Problems. Notes by Vijayasundaram Adimurthi, G., Tata Institute Lectures on Mathematics. Berlin-Heidelberg New York, Springer-Verlag & Bombay, Tata Institute of Fundamental Research 1980. XI, 240 S., DM 18,–. US $ 8.60. ISBN 3-540-08774-5

Kevorkian, J. / Cole, J. D., Perturbation Methods in Applied Mathematics. Applied Mathematical Sciences 34. Berlin-Heidelberg-New York, Springer-Verlag 1981. X, 558 S., 79 Abb., DM 88,–. US $ 41.90. ISBN 3-540-90507 3

Bühler, W. K., Gauss – A Biographical Study. Berlin-Heidelberg New York, Springer-Verlag 1981. VIII, 208 S., DM 39,–. US $ 17.80. ISBN 3-540-10662-6

Hua, L. K. / Wang, Y., Applications of Number Theory to Numerical Analysis. Berlin-Heidelberg-New York, Springer Verlag & Beijing; Science Press 1981. IX, 241 S., DM 78,–. US $ 37.20. ISBN 3-540-10382-1

NACHRICHTEN

Announcement

PLASTICITY TODAY

International Symposium on Current Trends and Results in Plasticity

Udine, June 1983

An International Symposium is proposed on the actual state of theories, methods and applications of plasticity. The emphasis is put on the last theoretical and experimental developments, current applications and expected relations with the bordering domains of mechanics and mathematics.

The Symposium will cover:

a) modelling of plastic behaviour in presence of various couplings;
b) mathematical and experimental methods suitable to deal with coupled problems involving thermal actions and damage, dynamics, cyclic response, localization of deformation, porosity and anisotropy inborn or that induced through hardening, softening or fissuration,
c) applications involving the phenomena invoked.

The Symposium will consist of:

a) a very limited number of general lectures intended to survey the contemporary plasticity and to stimulate specific border domains;
b) contributed papers within the scope of the Symposium.

The contributed papers are welcomed on various analytical, experimental and novel engineering aspects of plasticity. Persons interested in the meeting are requested to inform the CISM and to submit there by the end of December 1982 an information regarding possible participation and a summary of not less than 500 words of their proposed contribution.

The meeting relates to the Symposium on Plasticity in 1972 which resulted in publications of the volumes "Foundations of Plasticity" and "Problems of Plasticity". It is therefore purposely concerned with novel developments. Proceedings are planned to be published in a special volume.

The Symposium will be dedicated to the memory of Professor Wacław Olszak. Co-Chairmen of the Symposium are G. Bianchi and A. Sawczuk.

Address for correspondence:

CISM-Symposium PLASTICITY TODAY
Piazza Garibaldi, 18
33100 UDINE
Italy

FSC
www.fsc.org
MIX
Papier | Fördert gute Waldnutzung
FSC® C083411

Zeitfracht Medien GmbH
Ferdinand-Jühlke-Straße 7
99095 Erfurt, Deutschland
produktsicherheit@kolibri360.de